Dialysis Access Cases

Alexander S. Yevzlin • Arif Asif
Robert R. Redfield III • Gerald A. Beathard
Editors

Dialysis Access Cases

Practical Solutions to Clinical Challenges

 Springer

Editors
Alexander S. Yevzlin, MD, FASN
Professor of Medicine
University of Michigan
Ann Arbor, MI, USA

Robert R. Redfield III, MD
Assistant Professor of Transplant Surgery
Division of Transplant Surgery
Department of Surgery
University of Wisconsin
Madison, WI, USA

Arif Asif, MD, MHCM, FASN, FNKF
Professor of Medicine
Chairman: Department of Medicine
Jersey Shore University Medical Center
Seton Hall-Hackensack-Meridian
  School of Medicine
Neptune, NJ, USA

Gerald A. Beathard, MD, PhD
Co-Medical Director
Lifetime Vascular Access
Clinical Professor
University of Texas Medical Branch
Houston, TX, USA

ISBN 978-3-319-57498-1   ISBN 978-3-319-57500-1   (eBook)
DOI 10.1007/978-3-319-57500-1

Library of Congress Control Number: 2017943679

This Springer imprint is published by Springer Nature
The registered company is Springer International Publishing AG
The registered company address is: Gewerbestrasse 11, 6330 Cham, Switzerland

Preface

The practice of any procedural discipline is both a science and an art. In previous textbooks, edited by some of us, we tried to summarize the current state of hemodialysis access science. The purpose of this textbook, on the other hand, is to focus on the art of this medical discipline.

In this book the reader will find a multitude of cases, summarized by masters of the art of vascular access care, who articulate a broad, diverse, and creative vision of their practice. Clinical problems from routine access creation to advanced novel techniques are described in these pages. Thus, the purpose of this textbook is to educate the novice as well as to delight the expert.

Needless to say, we took great inspiration from our patients in putting this work together. Vascular access care requires repeated contact with the same patients on a regular basis. As a result, tremendously close bonds are formed. In some instances, we shared the chapters with our patients who "starred" in the cases. In all instances, we are deeply indebted to our patients for allowing us to use our minds, our hands, and our hearts to help them.

Ann Arbor, MI, USA   Alexander S. Yevzlin
Neptune, NJ, USA   Arif Asif
Madison, WI, USA   Robert R. Redfield III
Houston, TX, USA   Gerald A. Beathard

Contents

Contributors

Editors

Alexander S. Yevzlin, MD, FASN University of Michigan, Ann Arbor, MI, USA

Arif Asif, MD, MHCM, FASN, FNKF Department of Medicine, Jersey Shore University Medical Center, Seton Hall-Hackensack-Meridian School of Medicine, Neptune, NJ, USA

Robert R. Redfield III, MD Division of Transplant Surgery, Department of Surgery, University of Wisconsin School of Medicine and Public Health, Madison, WI, USA

Gerald A. Beathard, MD, PhD University of Texas Medical Branch, Houston, TX, USA

Authors

Anil K. Agarwal, MD Division of Nephrology, Department of Internal Medicine, The Ohio State University Wexner Medical Center, Columbus, Ohio, USA

Paul J. Foley, MD Department of Vascular and Endovascular Surgery, Hospital of the University of Pennsylvania, Philadelphia, PA, USA

Ali Gardezi, MBBS University of Wisconsin-Madison, Madison, WI, USA

Elliot I. Grodstein, MD Division of Transplant Surgery, Department of Surgery, University of Wisconsin School of Medicine and Public Health, Madison, WI, USA

Nabil J. Haddad, MD Division of Nephrology, Department of Internal Medicine, The Ohio State University Wexner Medical Cente, Columbus, OH, USA

Karl A. Illig, MD Division of Vascular Surgery, Department of Surgery, University of South Florida Morsani College of Medicine, Tampa, FL, USA

Venkat Kalapatapu, MD Perelman School of Medicine, University of Pennsylvania, Penn Presbyterian Hospital, Department of Surgery, Philadelphia, PA, USA

Muhammad Karim, MBBS Department of Medicine, University of Wisconsin-Madison, Madison, WI, USA

Kathleen M. Lamb, MD Hospital of the University of Pennsylvania, Department of Vascular and Endovascular Surgery, Philadelphia, PA, USA

Christopher M. Luty, MD University of Wisconsin-Madison, Department of Radiology, Vascular and Interventional Radiology, Madison, WI, USA

Jason W. Pinchot, MD Department of Radiology, Vascular an Interventional Radiology Section, University of Wisconsin School of Medicine and Public Health, Madison, WI, USA

Andre Ramdon, MBBS Hospital of the University of Pennsylvania, Department of Surgery, Philadelphia, PA, USA

Eduardo Rodriguez, MD Memorial Regional Hospital, Department of Vascular Surgery, Hollywood, FL, USA

Joel E. Rosenberg, MD Division of Nephrology, Department of Medicine, University of Wisconsin School of Medicine and Public Health, Madison, WI, USA

Paul A. Stahler, MD Division of Transplant Surgery, Department of Surgery, Hennepin County Medical Center, Minnesota, United States

Tushar J. Vachharajani, MD, FASN, FACP W.G. (Bill) Hefner VAMC, Salisbury, NC, USA

Part I
Arteriovenous Fistula

Chapter 1
Arteriovenous Graft-Arteriovenous Fistula Conversion (Secondary Arteriovenous Fistula Creation)

Elliot I. Grodstein and Robert R. Redfield III

Abbreviations

AV	Arteriovenous
FFCL	"Fistula First Catheter Last" Coalition
K/DOQI	Kidney Disease Outcomes Quality Initiative
sAVF	Secondary arteriovenous fistula

Case Presentation

A 65-year-old male has been dialyzed for 4 years via a left forearm looped brachio-basilic arteriovenous (AV) polytetrafluoroethylene graft. At the time of access creation, his basilic, median antebrachial, and cephalic veins in his forearm and upper arm were not suitable for construction of a primary AV fistula. Over the past few months, he has had difficulty achieving adequate dialysis flow rates during his Monday/Wednesday/Friday sessions. On exam, he is noted to have a pulse in his graft. He was referred to an interventional nephrologist who performed an angiogram demonstrating a venous anastomotic stenosis of greater than 50% of the luminal diameter (see Fig. 1.1). There was no upper arm or central venous stenosis present. An angioplasty was performed, and the patient was dialyzed for another 2 months until the graft thrombosed. Despite attempts to reestablish flow, the thrombosis could not be corrected. A right-sided internal jugular hemodialysis catheter

E.I. Grodstein, MD (✉) • R.R. Redfield III, MD
Division of Transplant Surgery, Department of Surgery, University of Wisconsin School of Medicine and Public Health, Madison, WI, USA
e-mail: grodstein@surgery.wisc.edu

© Springer International Publishing AG 2017
A.S. Yevzlin et al. (eds.), *Dialysis Access Cases*,
DOI 10.1007/978-3-319-57500-1_1

	DATE	DATE	DATE	DATE	DATE	DATE
Patient Name: _______________ **Staff Member Assigned:** _______________ **AV Graft Placement Date:** _______________						
1. Monthly Sleeves Up Exam Suitable outflow vein identified? If yes, proceed with next step. (Note: Suitable outflow vein not developed = Continue to monitor monthly for interventions per protocol)	Yes ☐ No ☐	Yes ☐ No ☐	Yes ☐ No ☐	Yes ☐ No ☐	Yes ☐ No ☐	Yes ☐ No ☐
2. Notify nephrologist. Obtain orders for either: 　☐ Fistulogram 　☐ Doppler Flow Study Results: _______________	Date Done:	Date Done:	Date Done:	Date Done:	Date Done:	Date Done:
3. If the test study is normal, cannulate the outflow vein with the venous needle for two consecutive treatments. (Note: If unable to successfully cannulate, continue to monitor monthly)	Date 1 Date 2	Date 1 Date 2	Date 1 Date 2	Date 1 Date 2	Date 1 Date 2	Date 1 Date 2
4. If cannulation successful, discuss plan with multidisciplinary team and patient. 　Plan A. ☐ Immediate conversion　　Plan B. ☐ Wait until access dysfunction evident 　Criteria for conversion: ☐ Impending graft thrombosis　☐ Graft thrombosis　☐ Other						
5. Surgery for secondary AVF access has been scheduled: 　Hospital / Access Center: _______________ 　Surgeon / Interventionalist: _______________	Date:	Date:	Date:	Date:	Date:	Date:

This educational item was produced through the AV Fistula First Breakthrough Initiative Coalition, sponsored by the Centers for Medicare and Medicaid Services (CMS), Department of Health and Human Services (DHHS). The content of this publication does not necessarily reflect the views or policies of the DHHS, nor does mention of trade names, commercial products, or organizations imply endorsement by the U.S. Government. The author(s) assume full responsibility for the accuracy and completeness of the ideas presented, and welcome any comments and experiences with this products.

Fig. 1.1 Sleeves Up Protocol checklist. Courtesy of End Stage Renal Disease National Coordinating Center

was placed. The patient was subsequently referred to a local vascular surgeon for reconstruction of peripheral access. Approximately 4 months later, a contralateral brachiocephalic AV fistula was constructed. The catheter was removed 2 months later, once the fistula matured. In the interim, he was hospitalized once with a methicillin-resistant *Staphylococcus aureus* bloodstream infection. The line was removed and replaced on the contralateral side. He was treated with vancomycin immediately after his hemodialysis sessions for 4 weeks.

Discussion

Unfortunately, as in the case above, AV graft failure is all too frequent, requiring subsequent catheter placementa. The National Kidney Foundation's Kidney Disease Outcomes Quality Initiative (K/DOQI) clinical practice guidelines prioritize the construction of autogenous AV fistulae in a distal-to-proximal fashion, always considering the preservation of more proximal surgical sites for future access construction. This emphasis is based on fistula's inherent higher primary and secondary patency rates and, with that, a lower need for interventions and, in select groups, longer patient survival. The K/DOQI guidelines, along with the "Fistula First Catheter Last" (FFCL) Coalition, seem to have changed practice patterns in the

United States [1]. From 1996 to 2007, autogenous fistula access use in the United States increased from 24% to 47%, while prosthetic graft use decreased from 58% to 28%. Still, however, 53% of patients are dialyzed via a central venous catheter or AV graft. To contextualize, in Japan, according to the collaborative international Dialysis Outcomes and Practice Patterns Study, this number is only 9% [2].

In most cases, as in the case above, AV graft failure is followed by a period of catheter placement prior to construction of new access. This cycle often repeats itself leading to unnecessary catheter days and high risks of central line-associated bloodstream infections. To avoid this, the FFCL has advocated for proactive construction of new AV fistulas prior to graft failures, so-called secondary AV fistulas (sAVF). The FFCL recommends nephrologists evaluate every AV graft patient for possible sAVF conversion [3, 4] and has put forth a convenient "Sleeves Up Protocol" to assist with this. The evaluation, which occurs briefly at bedside immediately prior to or after a dialysis session, helps identify suitable outflow vein for immediate conversion to an AVF. Every month, a practitioner should roll the patient's "sleeves up" exposing the entire arm up to the shoulder. Then, the upper arm should be lightly compressed to assess the caliber and prominence of the graft's venous outflow. If the primary outflow vein appears suitable for access, it should be cannulated with the venous dialysis access needle for two consecutive hemodialysis sessions. If there are no issues, a fistulogram or duplex ultrasound of the arm should be performed to confirm the vein's suitability and ensure patent venous drainage back to the right atrium. Assuming both of these tests go well, a prompt sAVF conversion plan should be made (see Fig. 1.1).

While AV grafts may spontaneously stop working, graft dysfunction is often predictable. Indicators of venous outflow stenosis include strong pulsatility and shortening or even absence of the diastolic phase of the thrill on exam. In severely obstructed grafts, there may be only a high-pitched thrill during systolic phase. Indicators of outflow problems on dialysis may include low flow rates, high venous pressures, or increased recirculation. There may be persistent post-dialysis bleeding. Other predictors of graft failure include requiring multiple interventions to maintain patency. Cumulative patency at 12 months of angioplastied grafts is around 30–50%, whereas it is only around 10–20% in thrombectomized grafts. Unfortunately, recent studies have shown no benefit to prophylactic treatment of graft stenosis detected on routine angiography. Thus, a preemptive sAVF conversion plan should be considered in patients with graft dysfunction requiring endovascular therapy. Similarly, as AV grafts have a greater rate of infection than autologous fistulae, sAVF is an attractive option to preserve access in patients requiring graft excision from recurrent infections. All of this forms the basis for the FFCL's recommendation that evaluation for sAVF conversion takes place no later than the first signs of AV graft failure. Meanwhile, the impetus is on the surgeon to perform the operation prior to a second intervention for graft stenosis or thrombosis. As such, the prudent surgeon should plan for sAVF conversion while placing an AV graft in a patient with initially unsuitable venous targets (see Fig. 1.2). In fact, in these cases, if a patient is being dialyzed via a catheter, an AV graft can be placed as a bridge to sAVF. Here, an immediate-access graft conduit (e.g., Flixene) is used to

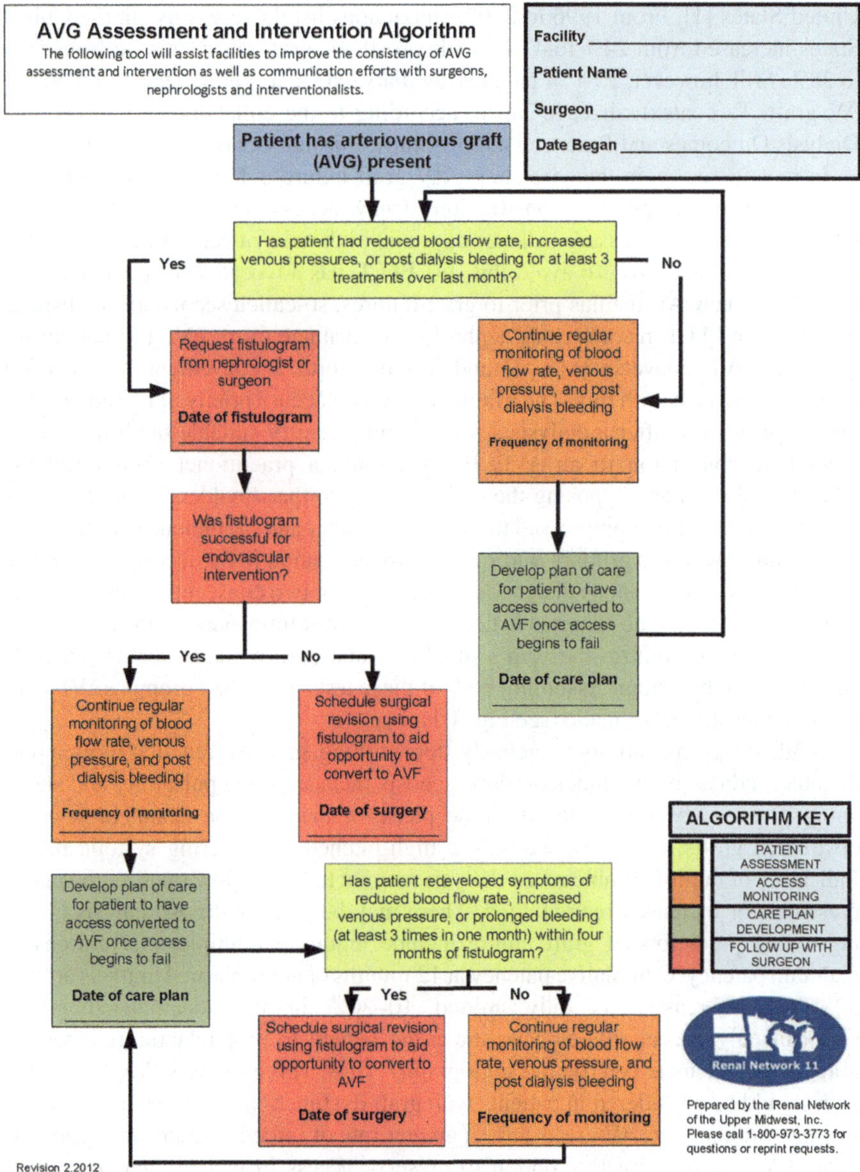

Fig. 1.2 AV graft assessment and intervention algorithm (Reprinted with permission from The Midwest Kidney Network)

allow prompt removal of the catheter. Months later, once the outflow becomes suitable, the graft can be excised and a sAVF constructed.

The classic sAVF, constructed using the main outflow vein from a graft, is termed a Type 1 sAVF. These fistulae can be constructed using below- or above-the-elbow

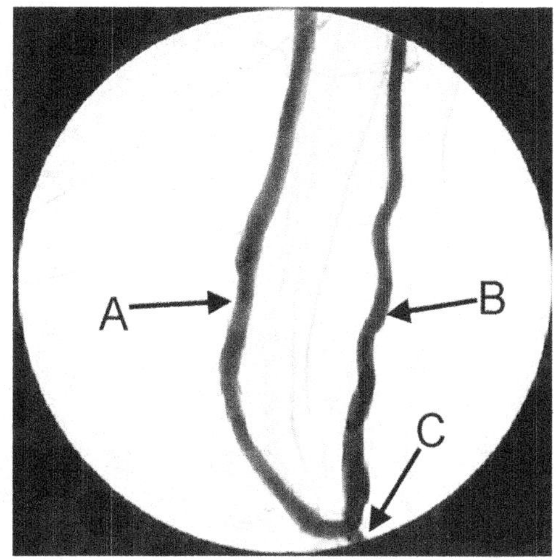

Fig. 1.3 Angiogram of upper arm veins draining a forearm loop graft: (**a**) basilic vein and (**b**) cephalic vein (**c**) venous anastomosis with lower arm AV graft (Reprinted with permission from Beathard [5])

veins following placement of a forearm AV graft. Duplex sonography and contrast venography are essential to both identify outflow veins of sufficient caliber and rule out central venous stenosis. Approximately 75% of patients with forearm AV grafts have vascular anatomy suitable for construction of a Type 1 sAVF (see Fig. 1.3). In the simplest scenarios, the sAVF can be made using arterialized basilic, cephalic, or brachial vein just distal to the previous venous anastomosis. Arterial inflow is typically provided by the brachial or proximal radial artery. Intraoperatively, the AVG is ligated and the outflow vein is mobilized and used for an anastomosis. To gain additional length, venous tributaries flowing into the suitable vein can be mobilized and used for an anastomosis. Frequently, after construction, Type 1 sAVFs are immediately available for dialysis access. A basilic or brachial vein can be transposed to a superficial position at the same time or as a staged approach. Even despite complete graft thrombosis, primary draining veins are often kept patent by tributaries, and a fistula can be constructed (see Fig. 1.4). In these settings, timely Type 1 sAVF construction can avert the need for catheter placement where it would otherwise be necessary.

In cases where the AV graft outflow is not amenable to Type 1 sAVF creation, vein mapping of the ipsilateral arm should be performed with the goal of finding other veins suitable for fistula creation. More frequently, these are patients with upper arm AV grafts, where only a small segment of primary outflow vein exists distally to the axilla. It remains important to still ensure that there is no central venous stenosis. As for termed Type 2 sAVFs, these are not dissimilar from standard primary AV fistulae and must be promptly constructed. The ipsilateral proximal radial artery will often have adequate inflow in an untouched surgical field. Venous targets may be more difficult to find however. Other than the median antebrachial or cephalic veins, in the setting of more proximal obstruction, forearm veins may be used in a retrograde fashion once valves are obliterated. Perforating veins may also be used. Overall, proximal radial artery fistulas have low rates of steal with excellent

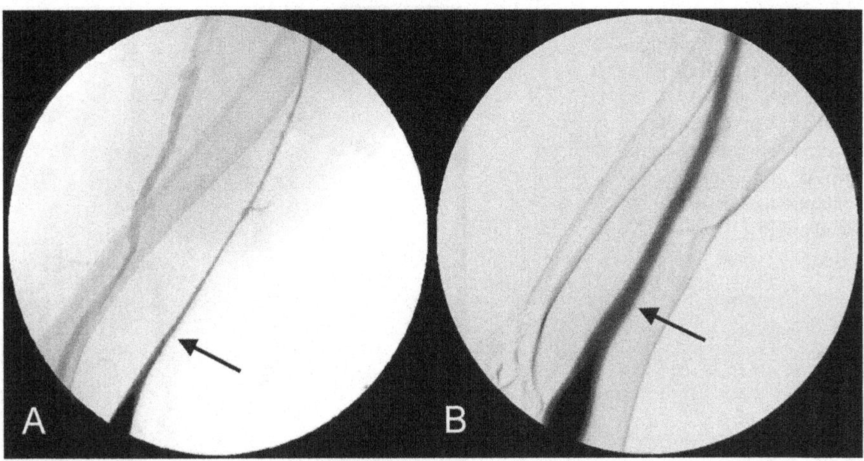

Fig. 1.4 Appearance of the left arm cephalic vein receiving drainage from a clotted forearm AV graft. The cephalic vein is indicated by the *arrow*. (**a**) Appearance when the graft is clotted with no flow or pressure. (**b**) Same vein with flow and pressure after graft was opened (Reprinted with permission from Beathard [5])

2-year secondary patency rates. If however there are no inflow or outflow options on the ipsilateral arm, the contralateral arm must be interrogated for de novo fistula creation. Unfortunately, whereas a failing AV graft will hopefully last long enough to allow for maturation, a failed AV graft will require interim dialysis catheter placement. Therefore, early recognition of graft dysfunction and construction of autologous access are a priority.

Pearls

- New autogenous access sites should be considered prior to AV graft failure to avert the need for hemodialysis catheter placement.
- Primary dialysis providers can evaluate patients for suitability for referral for secondary AV fistula formation quickly, during hemodialysis sessions.
- Secondary AV fistulas can frequently be constructed using the primary outflow of an AV graft, allowing immediate use for hemodialysis.

References

1. Park HS, et al. Comparison of outcomes with Arteriovenous fistula and arteriovenous graft for vascular access in hemodialysis: a prospective cohort Study. Am J Nephrol. 2016;43(2):120–8.

2. Becker B, Mitch W, Blake P, Obrador G, Collins A, Parekh R et al. NKF KDOQI Guidelines. http://www2.kidney.org/professionals/KDOQI/guideline_upHD_PD_VA/. Accessed 13 Apr 2016.
3. Ethier J, Mendelssohn DC, Elder SJ, Hasegawa T, Akizawa T, Akiba T, et al. Vascular access use and outcomes: an international perspective from the dialysis outcomes and practice patterns Study. Nephrol Dial Transplant. 2008;23(10):3219–26.
4. ESRD NCC. Secondary AVF placement in patients with AV grafts. http://esrdncc.org/ffcl/change-concepts/change-concept-6/. Accessed 13 Apr 2016.
5. Beathard GA. Interventionalist's role in identifying candidates for secondary fistulas. Semin Dial. 2004;17(3):233–6.

Chapter 2
Proximal Forearm Arteriovenous Fistula Creation

Venkat Kalapatapu and Andre Ramdon

Introduction

Worldwide greater than two million patients need renal replacement therapy. The aging population coupled with increasing need of dialysis access leaves surgeons gaining innovative ways to solve this crisis. Guidelines developed by the DOQI and Society of Vascular Surgery advised the creation of autologous fistula due to better long-term patency, fewer re-interventions, lower health-care cost, and low incidence of complications before use of grafts [1, 2]. Fistula first initiative uses distal radiocephalic and snuff box as the first choice, but this is impeded by lack of appropriately sized vessels for fistula creation and relatively high rates of non-maturations (8–40%).

Proximal forearm fistula has become grossly overlooked, likely due to the paucity of published literature, but is still a viable option. This preserves arm vessels for future use and has the theoretical advantage of reduced risk of steal syndrome, ischemic monomelic neuropathy, and high-output cardiac failure.

Proximal Radiocephalic Arteriovenous Fistula

Proximal radiocephalic arteriovenous fistula (pRCF) is an infrequently used option between the proximal radial artery and cephalic vein, first described in 1997 by Gracz et al. [3].

V. Kalapatapu, MD (⊠)
Perelman School of Medicine, University of Pennsylvania, Penn Presbyterian Hospital, Department of Surgery, Philadelphia, PA, USA
e-mail: Venkat.kalapatapu@uphs.upenn.edu

A. Ramdon, MBBS
Hospital of the University of Pennsylvania, Department of Surgery, Philadelphia, PA, USA

© Springer International Publishing AG 2017
A.S. Yevzlin et al. (eds.), *Dialysis Access Cases*,
DOI 10.1007/978-3-319-57500-1_2

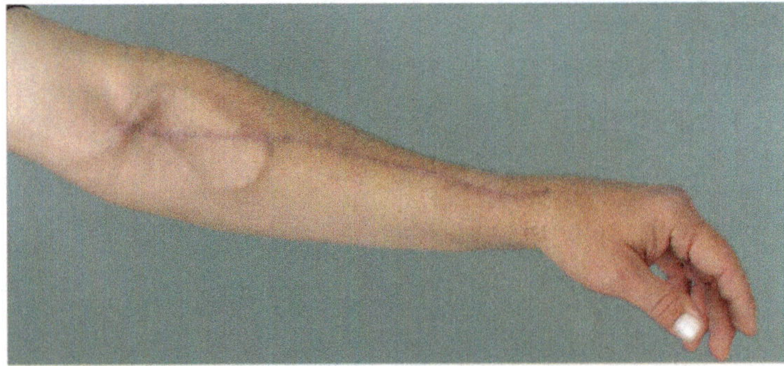

Fig. 2.1 Proximal radiocephalic fistula (the anastomosis can be to the radial artery at the elbow)

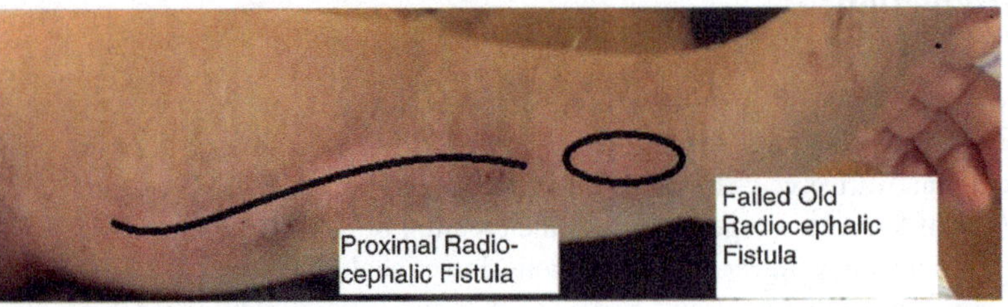

Fig. 2.2 Proximal radiocephalic fistula after failed distal radiocephalic fistula

This is an end-to-side construct through a 4–6 cm longitudinal incision commencing 1–2 cm below the antecubital crease and along the separation between the brachioradialis and flexor carpi radialis muscle. This allows mobilization of the cephalic vein and radial artery for an end-to-side anastomosis. This has a few options for venous outflow from the cephalic veins to the medial antecubital or a perforating vein. If the inflow through the radial artery is unfit, then the medial portion of the proximal incision can be extended to utilize the brachial artery for inflow.

The anastomosis is usually very deep and covered by muscles, which is thought to be protective. A large series of 105 patients reported a 91% primary patency after 11 months of follow-up [4]. A retrospective single-institution review of proximal vs. distal radiocephalic fistula revealed that patients are more likely to have had previous access (47% vs. 18%) and despite this have a low primary failure rate (32% vs. 59%). Cumulative pRCF vs. distal radiocephalic fistula patency was 92% vs. 86% at 1 year [5]. Proximal radiocephalic fistula is an attractive option for non-maturation distal radiocephalic fistula as the cephalic vein is likely more sizable distally and using the radial artery as inflow will limit the risk of steal and preserving the brachial inflow, thus, limiting steal and ischemic monomelic neuropathy (Figs. 2.1 and 2.2).

Basilic Vein Arteriovenous Fistulas in the Forearm

Forearm basilic vein fistula remains an underutilized option for fistula creation, even though advocated for by some authors on the basis of preservation of arm veins with patency rates greater than forearm grafts. The availability of literature with forearm basilic vein is rather scant, and the guidelines continue to exclude this as an option. Basilic veins can be anatomically deep in location making accessibility difficult but more likely preserved due to hidden nature and less susceptibility from multiple needle sticks or even previous access attempts. There are two constructs that can be utilized using either the radial or ulnar artery as the inflow. Both types are suggested to need transposition to allow for dialysis needle access. It is anxiety provoking to create an ulnar-basilic AVF (UBAVF) after a failed distal radiocephalic fistula due to the increased risk of distal ischemia reported to be 28% in one series. UBAVF has much higher failure rates and longer maturity times compared to distal radiocephalic fistula with 1-year patency rates that range from 42% to 70% and a secondary patency rate of 53% [6]. Complication of hand ischemia is 0.4% in one pooled analysis.

Transposed radio-basilic fistula (tRBF) is gaining favor and is particularly attractive after failure of distal radiocephalic fistula with a reported 1-year patency rate as high as 93% in a small series of 30 patients (mostly after a thrombosed cephalic vein) [7–9]. Patency rates are non-inferior to arteriovenous grafts but more importantly without the infectious risk. Additionally, if not matured enough to be used for dialysis, it will contribute to the increased size of arm veins and hence extrapolates to improved outcomes of a more proximal fistula patency at a later date. One study comparing tRBF vs. arteriovenous graft proves fistula first is better with reported patency periods of 16.9 vs. 12.6 months with primary-assisted patency at 1 year 79% and 75%, respectively [9, 10]. Compared to distal radiocephalic fistula, primary patency rates are lower at 1 year (40–54%) and with maturation failure as high as 14% [7, 10]. Shintaro and Natario et al. suggested that low initial patency could be improved with intense observation and surveillance with early introduction of balloon angioplasty to increase as much as 77% [7, 11].

Procedurally, tRBF is more difficult with longer operative time but still feasible under local anesthesia. Preoperative duplex ultrasound is important in patient selection and planning of these fistulas. Technique is key: skin sparing with three to four separate incisions for harvest or long elbow-to-wrist incision with a counter-incision over the approximate radial artery after tunneling of available vein. The basilic vein usually runs a little far from the arteries; hence, usually the best positioning during harvesting is flexion at the elbow with forearm supination. General principles of harvesting apply with special care not to injure the vein. The basilic vein after ligation of the side branches forms a high-resistance conduit which is prone to thrombosis [8]. Once the vein is harvested, great care is taken to gently angio-dilate. Some authors prefer to use a 3/4 Fogarty catheter. Meticulous tunneling then allows for subcutaneous access and anastomosis to the radial artery. This moves the vessel away from its native course which can be deep and restrictive with scar tissue formation and healing. Anastomosis is created in an end-to-side construct with

Glowinski et al. suggesting 6 mm as being better than a larger diameter [8]. This can be performed in the forearm from the brachial to the distal radial depending on the suitability with maximizing the entire vein by looping as is needed. Outcomes are dependent on vein size with 3.5 mm vein yielding patency of 93/78/55% at 1, 2, and 3 years, respectively. Use of 2.5 mm veins yields a 1-year patency of 54% [7]. Duplex ultrasound should also demonstrate good inflow with radial artery diameter of >2.5 mm. Silva classified anatomic variants of the basilic vein into three types [9]. Type A (15%) vein is close to the radial artery, and a single incision is needed for harvest and creating anastomosis. Type B (33%) vein is located dorsally, and type C (52%) vein is more volar. Both B and C require separate incisions for harvest and anastomosis, but all will need superficialization for the normal deep position.

Conclusion

Proximal forearm fistula remains an untapped resource for fistula creation, which has escaped the guidelines but is with acceptable patency rates and preservation of arm veins for future use. Additionally, this offers a theoretical reduced risk of steal syndrome, ischemic monomelic, and high-output cardiac failure. This requires a skilled and highly experienced team of surgeons, nephrologists, and dialysis nurses to ensure the success of these accesses. More studies are encouraged to continue for the improvement of these unique proximal forearm fistulas.

References

1. Depner T, Daugirdas JT. KDOQI clinical practice guidelines and clinical practice recommendations for vascular access. Am J Kidney Dis. 2006;48(Suppl. 1):S176–322.
2. Sidawy AN, Spergel LM, Besarab A, Allon M, Jennings WC, Padberg FT Jr, Murad MH, Montori VM, O'Hare AM, Calligaro KD, Macsata RA, Lumsden AB, Ascher E. The Society for Vascular Surgery: clinical practice guidelines for the surgical placement and maintenance of arteriovenous hemodialysis access. J Vasc Surg. 2008;48(5 Suppl):2S–25S.
3. Gracz KC, Ing TS, Soung LS, Armbruster KFW, Seim SK, Merkel FK. Proximal forearm fistula for maintenance hemodialysis. Kidney Int. 1977;11:71–4.
4. Jennings WC. Creating arteriovenous fistulas in 132 consecutive patients: exploiting the proximal radial artery arteriovenous fistula: reliable, safe, and simple forearm and upper arm hemodialysis access. Arch Surg. 2006;141:27–32.
5. Bhalodia R, Allon M, Hawxby AM, Maya ID. Comparison of radiocephalic fistulas placed in the proximal forearm and in the wrist. Semin Dial. 2011;24(3):355–7.
6. Al Shakarchi J, Khawaja A, Cassidy D, Houston JG, Inston N. Efficacy of the ulnar-basilic arteriovenous fistula for hemodialysis: a systematic review. Ann Vasc Surg. 2016;32:1–4.
7. Kumano S, Itatani K, Shiota J, Gojo S, Izumi N, Kasahara H, Homma Y, Tagawa H. Strategies for the creation and maintenance of reconstructed arteriovenous fistulas using the forearm basilic vein. Ther Apher Dial. 2013;17(5):504–9.
8. Glowinski J, Glowinska I, Malyszko J, Gacko M. Basilic vein transposition in the forearm for secondary arteriovenous fistula. Angiology. 2014;65(4):330–2.

9. Silva MB Jr, Hobson RW 2nd, Pappas PJ, Haser PB, Araki CT, Goldberg MC, Jamil Z, Padberg FT Jr. Vein transposition in the forearm for autogenous hemodialysis access. J Vasc Surg. 1997;26(6):981–6.
10. Son HJ, Min SK, Min SI, Park YJ, Ha J, Kim SJ. Evaluation of the efficacy of the forearm basilic vein transposition arteriovenous fistula. J Vasc Surg. 2010;51(3):667–72.
11. Natário A, Turmel-Rodrigues L, Fodil-Cherif M, Brillet G, Girault-Lataste A, Dumont G, Mouton A. Endovascular treatment of immature, dysfunctional and thrombosed forearm autogenous ulnar-basilic and radial-basilic fistulas for haemodialysis. Nephrol Dial Transplant. 2010;25(2):532–8.

Chapter 3
Fistula with Stenosis of Feeding Artery

Gerald A. Beathard

Case Presentation

The patient is a 75-year-old female with a radial cephalic fistula (AVF) in her left arm. The fistula was created 6 months ago but has not matured adequately for dialysis use. The patient was sent to the vascular access center for evaluation. Physical examination revealed that the fistula was not visible. A soft thrill and bruit, systolic only, were present over the anastomosis. The pulse augmentation was poor (3/10).

The AVF was cannulated in an upstream direction, and a sheath was inserted. A retrograde arteriogram of the anastomosis and lower radial artery was performed. This showed a focus of marked stenosis approximately 3 cm above the anastomosis (Fig. 3.1). This lesion was dilated using a 4 × 4 angioplasty balloon with good results (Fig. 3.2). Blood flow in the fistula was improved. Physical examination was repeated and showed improvement of the thrill and bruit at the anastomosis. Pulse augmentation was improved (5/10) but was not optimal. A guidewire was inserted, passed across the arterial anastomosis, and advanced up to the level of the subclavian artery. An antegrade arteriogram was performed which showed multiple areas of stenosis in the proximal radial artery just distal to the bifurcation (Fig. 3.3). A 5 × 4 angioplasty balloon was inserted over the guidewire and advanced up to the level of the stenoses. These were dilated with good result (Fig. 3.4).

Following this procedure, the radial cephalic AVF was visible to the level of the elbow. Blood flow in the fistula was good. The thrill and bruit at the anastomosis were systolic and diastolic and of good quality. Pulse augmentation improved to an optimal level (10/10). The AVF was used successfully for dialysis.

G.A. Beathard, MD, PhD (✉)
Lifetime Vascular Access, University of Texas Medical Branch,
5135 Holly Terrace, Houston, TX 77056, USA
e-mail: gbeathard@msn.com

© Springer International Publishing AG 2017
A.S. Yevzlin et al. (eds.), *Dialysis Access Cases*,
DOI 10.1007/978-3-319-57500-1_3

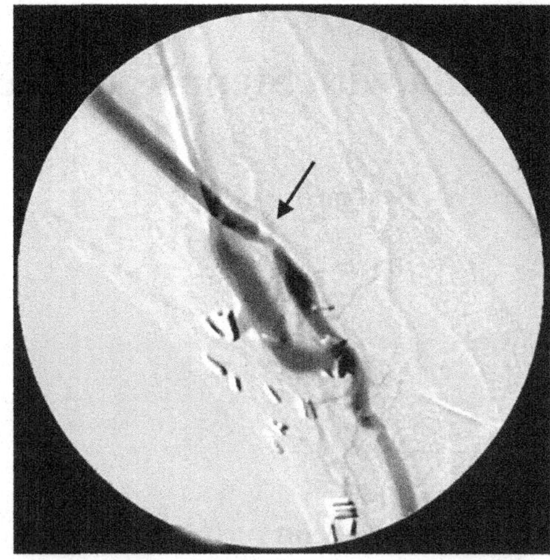

Fig. 3.1 Angiogram showing just anastomotic fistula and artery (Note the stenotic area in the artery approximately 3 cm above the anastomosis (*arrow*))

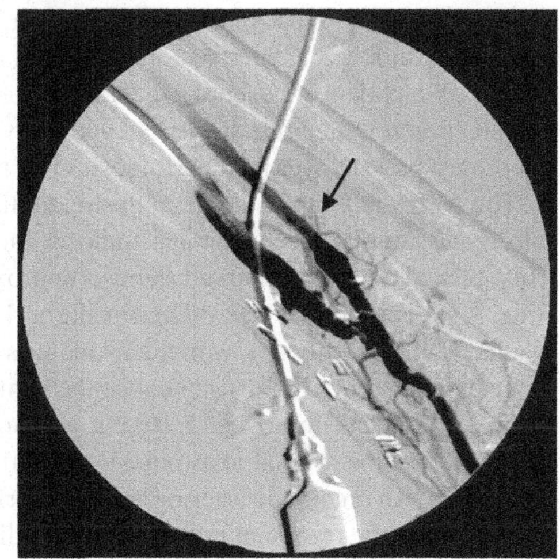

Fig. 3.2 Appearance of artery following treatment with 4 × 4 angioplasty balloon (*arrow* indicates site of previous lesion)

Discussion

The dialysis vascular access should be thought of as a complete circuit starting and ending with the heart. The venous side, the AVF and its draining veins, represents only one-half of the circuit; the other half is arterial. Lesions in this region adversely affect inflow. These may be within any of the arteries that ultimately lead to the access, or they can affect the arterial anastomosis, which is considered to be the arterial component of the access itself. Frequently these two types of lesions are reported together with juxta-anastomotic lesions as inflow stenosis. These lesions can lead to decreased blood flow in the access, which can result in inadequate dialysis if the AVF is functional and failure to mature in a newly created AVF.

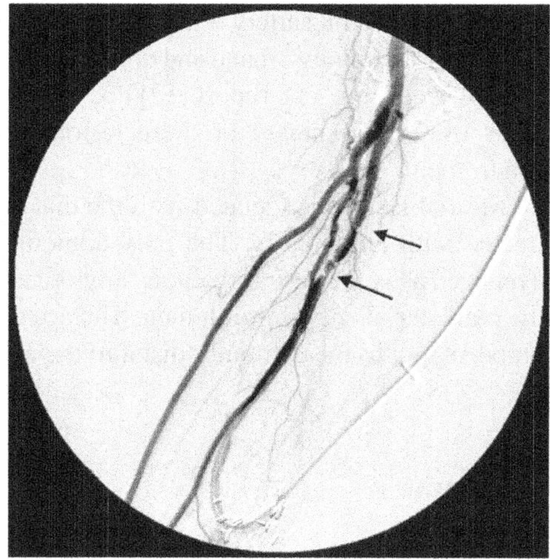

Fig. 3.3 Arteriogram showing bifurcation and proximal radial artery. Site of stenotic lesions indicated by *arrows*

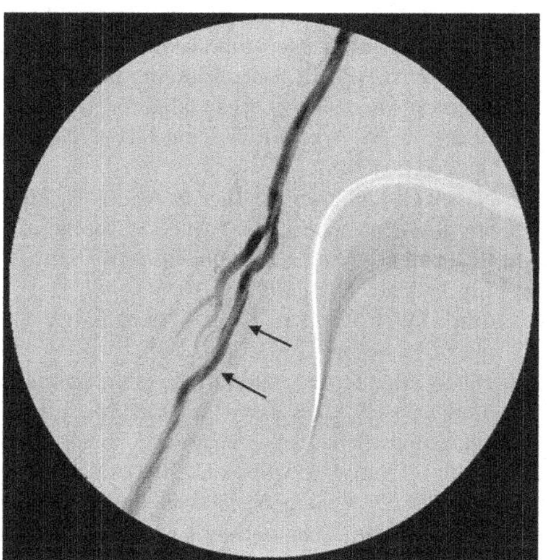

Fig. 3.4 Appearance of artery following treatment with 5 × 4 angioplasty balloon (*arrows* indicate sites of previous lesions)

In a report dealing with a cohort of 101 dysfunctional AVF cases [1], 8% were found to have lesions in the feeding artery, and 21% had stenosis of the arterial anastomosis. There was a higher incidence of inflow stenosis for forearm as compared to upper arm AVFs. Others [2, 3] have reported the incidence of arterial stenosis in these cases at 6–18%.

Inflow lesions are the most common cause of failure of a newly created AVF to mature. These lesions result in decreased fistula blood flow leading to problems of maturation and often early thrombosis. The most common inflow lesion resulting in failure of an AVF to mature is stenosis of the juxta-anastomotic segment of the AVF [2, 4–11]. This makes it the most common lesions associated with this problem; however, stenosis of the inflow artery also occurs and has a similar effect.

When the feeding artery is affected, the patient can also develop ischemic problems in the distal extremity—hand and digits. In an evaluation of 12 patients with symptoms of steal syndrome, one report [12] documented arterial stenotic lesions in ten of the cases (83%). Treatment of these lesions resulted in resolution of the ischemic syndrome.

Arterial lesions associated with the dialysis vascular access are generally easily treated with angioplasty. The only admonition is that the balloon should not be oversized as is the case for venous angioplasty. It is important that it be based upon the diameter of the arterial lumen. The artery also has a vital function in providing blood supply to the extremity distal to the anastomosis.

References

1. Asif A, Gadalean FN, Merrill D, et al. Inflow stenosis in arteriovenous fistulas and grafts: a multicenter, prospective study. Kidney Int. 2005;67:1986.
2. Turmel-Rodrigues L, Mouton A, Birmele B, et al. Salvage of immature forearm fistulas for haemodialysis by interventional radiology. Nephrol Dial Transplant. 2001;16:2365.
3. Romero A, Polo JR, Garcia Morato E, et al. Salvage of angioaccess after late thrombosis of radiocephalic fistulas for hemodialysis. Int Surg. 1986;71:122.
4. Beathard GA, Arnold P, Jackson J, et al. Aggressive treatment of early fistula failure. Kidney Int. 2003;64:1487.
5. Nassar GM, Nguyen B, Rhee E, Achkar K. Endovascular treatment of the "failing to mature" arteriovenous fistula. Clin J Am Soc Nephrol. 2006;1:275.
6. Beathard GA, Settle SM, Shields MW. Salvage of the nonfunctioning arteriovenous fistula. Am J Kidney Dis. 1999;33:910.
7. Clark TW, Cohen RA, Kwak A, et al. Salvage of nonmaturing native fistulas by using angioplasty. Radiology. 2007;242:286.
8. Falk A. Maintenance and salvage of arteriovenous fistulas. J Vasc Interv Radiol. 2006;17:807.
9. Manninen HI, Kaukanen E, Makinen K, Karhapaa P. Endovascular salvage of nonmaturing autogenous hemodialysis fistulas: comparison with endovascular therapy of failing mature fistulas. J Vasc Interv Radiol. 2008;19:870.
10. Shin SW, Do YS, Choo SW, et al. Salvage of immature arteriovenous fistulas with percutaneous transluminal angioplasty. Cardiovasc Intervent Radiol. 2005;28:434.
11. Badero OJ, Salifu MO, Wasse H, Work J. Frequency of swing-segment stenosis in referred dialysis patients with angiographically documented lesions. Am J Kidney Dis. 2008;51:93.
12. Asif A, Leon C, Merrill D, et al. Arterial steal syndrome: a modest proposal for an old paradigm. Am J Kidney Dis. 2006;48:88.

Chapter 4
Juxta-anastomotic Stenosis of Brachial-Cephalic Fistula

Gerald A. Beathard

Case Presentation

The patient was a 48-year-old male who had been on dialysis for 1 ½ years using a left arm brachial-cephalic arteriovenous fistula (AVF). The nurses at the dialysis facility reported that they had had increasing difficulty cannulating the AVF. The patient was referred to the dialysis access center with a clinical diagnosis of thrombosis.

This examination of the AVF revealed dialysis cannulation sites over the upper and lower AVF. A hyper-pulsatile segment was present which included the anastomosis and approximately 4 cm of the fistula. A thrill was not palpable in this area. The fistula above this point was not palpable. A bruit was not palpable. Evaluation using Doppler ultrasound revealed a point of severe stenosis in the juxta-anastomotic region with total obstruction approximately 4 cm above the anastomosis. No thrombus was evident within the AVF.

After the patient's arm was prepped and draped, the brachial artery was cannulated proximal to the anastomosis in a retrograde direction, and the angiogram was performed (Fig. 4.1). This showed severe juxta-anastomotic stenosis (Fig. 4.1b) with complete obstruction more proximally (Fig. 4.1a). Several collaterals were present which had blood flow. No thrombus was evident. An attempt was made to pass a guidewire through the fistula. The guidewire passed through the point of apparent obstruction and continued up through the cephalic arch to the central veins (Figs. 4.2, 4.3, 4.4, 4.5, and 4.6). The entire cephalic vein from the arch down to the anastomosis was patent without problems.

The patient was returned to the dialyzer successfully.

G.A. Beathard, MD, PhD (✉)
Lifetime Vascular Access, University of Texas Medical Branch,
5135 Holly Terrace, Houston, TX 77056, USA
e-mail: gbeathard@msn.com

© Springer International Publishing AG 2017
A.S. Yevzlin et al. (eds.), *Dialysis Access Cases*,
DOI 10.1007/978-3-319-57500-1_4

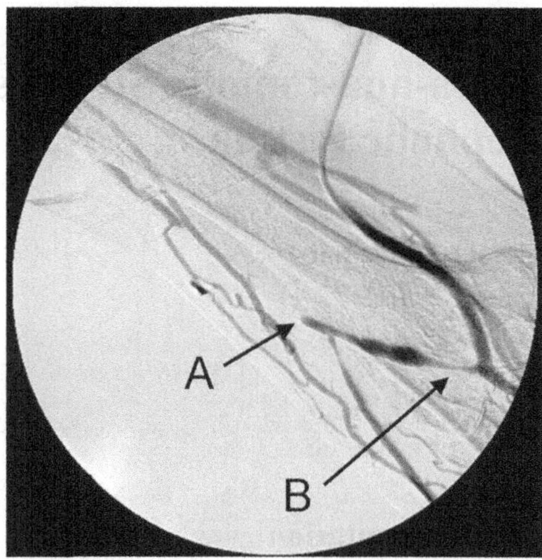

Fig. 4.1 Initial angiogram from brachial artery cannulation. *A* point of apparent total occlusion, *B* juxta- anastomotic stenosis

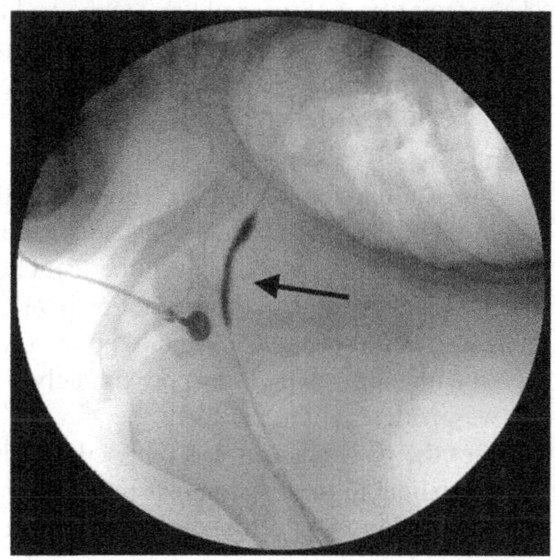

Fig. 4.2 Angioplasty of upper cephalic vein; *arrow* indicates angioplasty balloon

Discussion of the Case

A clinical diagnosis of AVF thrombosis has a very limited implication as it relates to treatment. Within this diagnosis is a relatively broad spectrum of presentations ranging from no thrombus present, as in this case, to the AVF that is markedly dilating, ectatic, and filled with a large volume of thrombus. Whereas a relatively simple algorithm can be successfully utilized for the treatment of virtually any thrombosed arteriovenous graft (AVG), this approach is inadequate for the treatment of a thrombosed AVF. This is due to the fact that an AVF behaves quite differently from an AVG. First, it can tolerate much lower blood flow rates without clotting. Because of this, an AVF clots later in the progression of a stenotic lesion that slows blood flow.

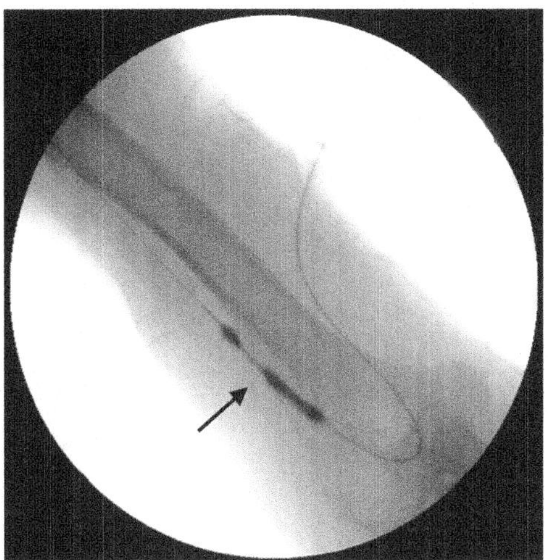

Fig. 4.3 Angioplasty of lower cephalic vein; *arrow* indicates angioplasty balloon

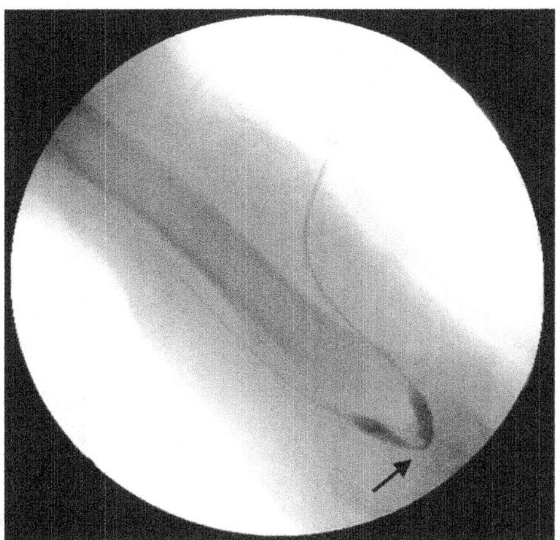

Fig. 4.4 Angioplasty of anastomosis; *arrow* indicates angioplasty balloon

The lesion has longer to develop and therefore is often much more severe than what is seen with an AVG. Second, an AVF generally develops collaterals that allow for a continuation of blood flow even after the lesion in the main body of the AVF has become totally obstructive as was seen in this case.

The broad spectrum represented by the diagnosis of AVF thrombosis requires individualization [1]. The interventionalist must evaluate the individual situation that is presented and determine what needs to be done to salvage the access. Another point illustrated by this case is the fact that very often a guidewire can be advanced through a lesion that appears to represent a total obstruction. One should not conclude that such a lesion actually will not permit a guidewire passage until an attempt has been made.

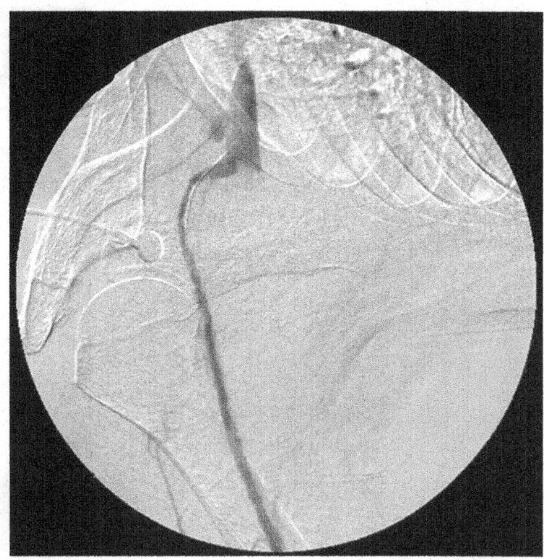

Fig. 4.5 Post-angioplasty angiogram of upper cephalic vein

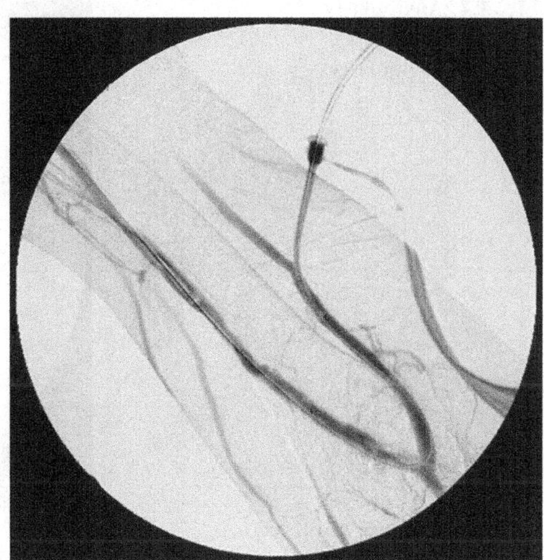

Fig. 4.6 Post-angioplasty angiogram of lower portion of fistula

Although most interventional nephrologists tend to approach a dysfunctional AVF through a venous cannulation, not all cases can be treated in this manner. As demonstrated in this case, an arterial approach is sometimes necessary.

Reference

1. Haage P, Vorwerk D, Wildberger JE, et al. Percutaneous treatment of thrombosed primary arteriovenous hemodialysis access fistulae. Kidney Int. 2000;57:1169.

Chapter 5
Fistula with Diffuse Venous Stenosis

Gerald A. Beathard

Case Presentation

The patient was a 48-year-old female with a history of chronic hypertension. A brachial-cephalic fistula was created in the patient's left arm without the benefit of vascular mapping. The fistula had failed to develop 10 weeks after creation. The patient was referred to the dialysis access center for evaluation. Physical examination revealed the presence of a surgical scar just above the crease of the elbow. The anastomosis was palpable and was felt to be hyper-pulsatile. The thrill and bruit were high pitched and systolic only. The upper arm cephalic vein was not palpable. Blood flow in the brachial artery was measured at 280 mL/min.

After the patient's arm was prepped and draped, the brachial artery was cannulated below the anastomosis in a retrograde direction. An angiogram was performed which showed a diffusely stenotic cephalic vein (Figs. 5.1 and 5.2). After passing a guidewire, the entire cephalic vein down to and including the arterial anastomosis was dilated with a 6 × 10 angioplasty balloon. Treatment was started approximately progressed distally (Figs. 5.3 and 5.4). The inflow into the fistula was manually occluded during each of the dilatations; occlusion was maintained until the angioplasty balloon was completely deflated. A follow-up angiogram showed a good result although there were a number of intimal tears throughout the entire length of the vein.

The patient was seen in follow-up 2 weeks following the angioplasty procedure. The fistula was visible and palpable throughout its entire length. It collapsed with arm elevation. The thrill and bruit at the anastomosis were continuous and low pitched. Ultrasound evaluation showed a diameter of 6 mm with blood flow of 820 mL/min.

G.A. Beathard, MD, PhD (⊠)
Lifetime Vascular Access, University of Texas Medical Branch,
5135 Holly Terrace, Houston, TX 77056, USA
e-mail: gbeathard@msn.com

© Springer International Publishing AG 2017
A.S. Yevzlin et al. (eds.), *Dialysis Access Cases*,
DOI 10.1007/978-3-319-57500-1_5

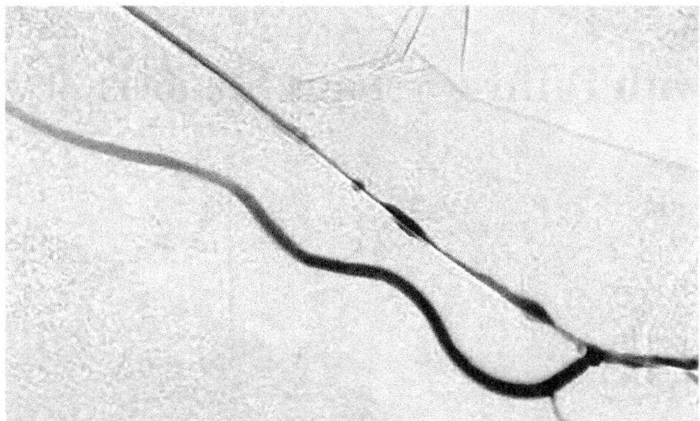

Fig. 5.1 Angiogram of lower brachial-cephalic fistula showing diffusion venous stenosis

Fig. 5.2 Angiogram of
upper brachial-cephalic
fistula showing diffusion
venous stenosis

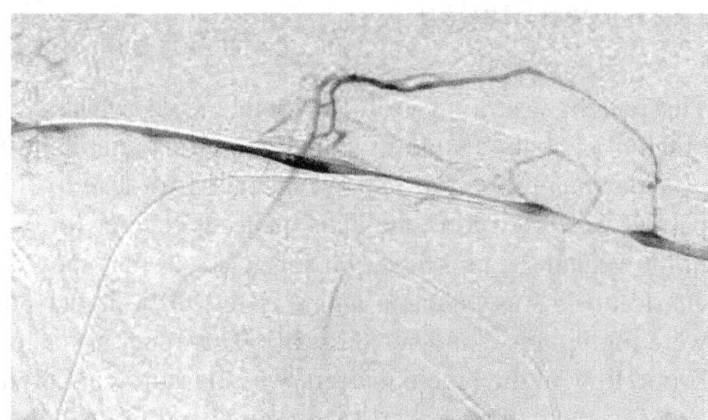

Fig. 5.3 Angioplasty of
upper fistula with 10×6
angioplasty balloon

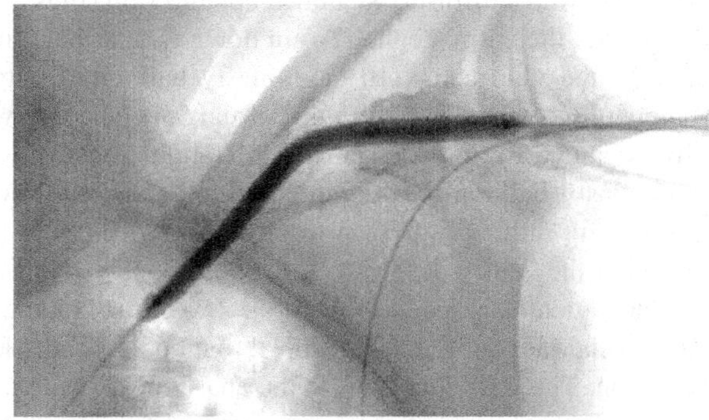

Discussion

The procedure that has been used to treat these cases is referred to as balloon-assisted maturation (BAM). In some cases, an AVF has been purposefully created using a small vessel in anticipation of performing this procedure in order to obtain the requisite size necessary for the AVF to be functional [1].

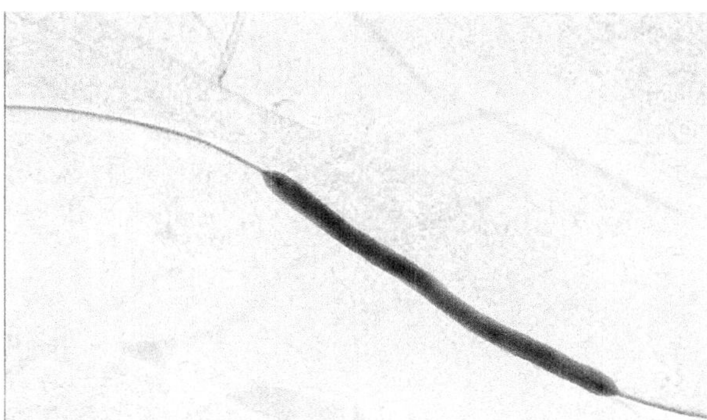

Fig. 5.4 Angioplasty of lower fistula with 10 × 6 angioplasty balloon

BAM is an aggressive approach to AVF maturation failure in which repeated long-segment angioplasty is used to sequentially dilate that portion of the vein destined to become the AVF. This procedure has been around for a long time; however, more recently the approach has been more aggressively applied. It appears that the use of BAM is becoming increasingly popular, despite few evidence-based studies and no randomized prospective trials. In some instances, the vein has actually been converted into what is essentially a collagen tube. In addition, there have been reports of performing this technique intraoperatively on both veins and arteries in an attempt to create an AVF in a patient with small vessels [2, 3].

The angioplasty procedure employed in order to achieve optimal BAM results is often a multistep process [4, 5]. This generally involves taking a relatively small vessel, 2–3 mm in diameter, up to 6 mm in the first stage followed by sequential treatments at 2–4-week intervals using a slightly larger angioplasty balloon until the desired endpoint is reached. The major complication associated with this procedure has been venous rupture. It has been postulated that this is actually associated with an endothelial tear that occurs at the shoulder of the balloon [5]. Intraluminal blood pressure working in concert with luminal obstruction by the partially deflated balloon (or any other type of obstruction in the lumen) results in extravasation (Fig. 5.5). In order to minimize this effect, several things have been recommended [4, 5].

First has been the use of a long angioplasty balloon, 10 cm in length. This serves to decrease the number of inflations that must be performed in order to dilate a long section of vein and therefore decreases the opportunity for this phenomenon rupture to occur.

Second, it has been recommended that arterial inflow be manually obstructed during the entire process from balloon dilatation through complete balloon deflation. This prevents the surge in intraluminal pressure that can potentially occur against the luminal obstruction of a partially inflated balloon.

Third, it is important that angioplasty of a long segment of vein proceeds from proximal to distal. Actually, this is the recommended progression for all angioplasty procedures. It is important that the outflow be unobstructed in order to

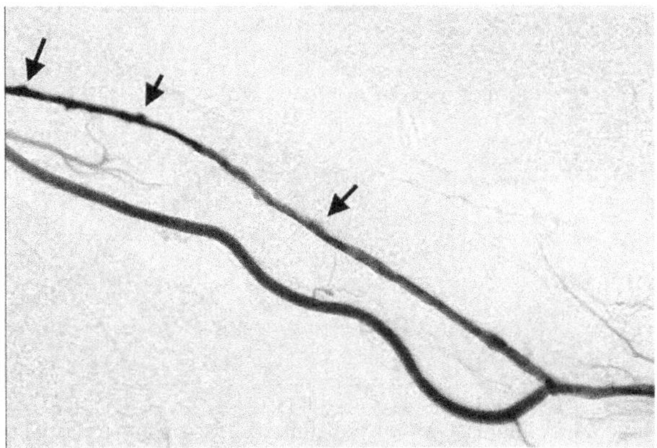

Fig. 5.5 Post-angioplasty angiogram showing dilatation of vein with multiple intimal tears (*arrows*)

decrease the chances of vein rupture and minimize the degree of extravasation that might follow such an event. At times, this may require both antegrade and retrograde cannulations.

Fourth, although there is no need for the use of heparin when doing a dialysis access angioplasty, it is felt that the avoidance of heparin when doing the BAM is even more important [4].

In one study [5], 122 patients with a non-maturing AVF were treated. Technical success was achieved in 118 of these cases. In this report, the AVFs were divided into two groups – class 1 consisted of those that were 6–8 mm in diameter but were more than 6 mm deep, and class 2 consisted of those that were 2–5 mm in diameter. BAM was used in the class 1 cases in an attempt to make the vessel large enough that it could be cannulated in spite of its depth. Follow-up was available on 109 of the successful cases. The number of procedures required to attain the treatment goal for class 1 and class 2 cases was 1.6 and 2.6, respectively. This was achieved in 5 and 7 weeks, respectively. The primary patency for class 1 and class 2 cases was 17% and 39% at 6 months, respectively. Secondary patency for the two groups was 72 and 77% at 12 months, 53 and 61% at 24 months, and 42 and 32% at 36 months, respectively.

In another study [4], 373 patients were treated with the BAM procedure. The initial angioplasty in the sequence was performed 2–4 weeks after the creation of the access. A 9.9% technical failure rate was experienced. These patients required a mean of 2.7 sequential dilatation sessions to reach the BAM endpoints.

The major complication associated with the BAM procedure has been vein rupture with hematoma formation. One study [5] reported that in their series, minor, self-limited venous rupture of some degree occurred in all of the patients. In 5% of the cases, significant venous rupture with extravasation did occur. Most of these were easily controlled with manual compression and balloon tamponade.

In another study [4], complications resulting in loss of the access occurred in 37 of 373 patients who received 1019 sequential angioplasty treatments. These consisted of 34 cases in which the AVF thrombosed, 1 case with an anastomotic rupture, and 2 vein ruptures severe enough to result in access loss.

In a study in which BAM was performed with ultrasound guidance [6], a complication rate of 54% in upper arm AVFs and 67% of those in the forearm was reported. In the upper arm ($n = 102$), the complications consisted of 35 hematomas, 14 vein ruptures, 2 cases of venous spasm, and 1 thrombosis. In the forearm ($n = 234$), there were 101 hematomas, 18 vein ruptures, 24 instances of venous spasm, and 4 thromboses. While this series demonstrates the types of complications that might be encountered, the increased incidence in comparison to other studies raises the question of the advisability of doing this procedure with only ultrasound guidance.

In an earlier publication [7], the same group reported a 53% incidence of the development of a hematoma within the wall of the vessel, something that would not be apparent without ultrasound imaging. They pointed out that this did not seem to affect flow and theorized that this event was an indication of a successful procedure, which had resulted in optimal vessel expansion.

References

1. Kotoda A, Akimoto T, Sugase T, et al. Balloon-assisted creation and maturation of an arterio-venous fistula in a patient with small-caliber vasculature. Case Rep Nephrol Urol. 2012;2:65.
2. De Marco Garcia LP, Davila-Santini LR, Feng Q, et al. Primary balloon angioplasty plus balloon angioplasty maturation to upgrade small-caliber veins (<3 mm) for arteriovenous fistulas. J Vasc Surg. 2010;52:139.
3. Veroux P, Giaquinta A, Tallarita T, et al. Primary balloon angioplasty of small (</=2 mm) cephalic veins improves primary patency of arteriovenous fistulae and decreases reintervention rates. J Vasc Surg. 2013;57:131.
4. Chawla A, DiRaimo R, Panetta TF. Balloon angioplasty to facilitate autogenous arteriovenous access maturation: a new paradigm for upgrading small-caliber veins, improved function, and surveillance. Semin Vasc Surg. 2011;24:82.
5. Miller GA, Goel N, Khariton A, et al. Aggressive approach to salvage non-maturing arteriovenous fistulae: a retrospective study with follow-up. J Vasc Access. 2009;10:183.
6. Derderian T, Hingorani A, Boniviscage P, et al. Acute Complications After Balloon Assisted Maturation. Ann Vasc Surg. 2014;28(5):1275–9.
7. DerDerian T, Hingorani A, Ascher E, et al. To BAM or not to BAM?: A closer look at balloon-assisted maturation. Ann Vasc Surg. 2013;27:104.

Chapter 6
Balloon-Guidewire Entrapment

Gerald A. Beathard

Case Presentation

A 53-year-old diabetic female dialysis patient was referred to the dialysis access center for non-maturation of a left radial-cephalic arteriovenous fistula (AVF). On physical examination of the AVF, approximately 3–4 mm of the vein was visible at the anastomosis which was hyper-pulsatile. Above this point, the cephalic vein was not palpable. A thrill was palpable at the anastomosis, which was systolic only. Examination of the AVF with Doppler ultrasound revealed the presence of juxta-anastomotic stenosis.

After the patient was prepped and draped, the AVF was cannulated immediately above the anastomosis in the venous segment that was visible. A guidewire was passed up to the level of the central veins. A 6 × 4 angioplasty balloon was advanced over the guidewire up to the upper forearm where it was inflated to the point of full effacement, but additional pressure was not applied. The inflated angioplasty balloon was then cannulated using a micropuncture needle (Fig. 6.1). As soon as a clear-fluid flashback appeared at the needle hub, the guidewire was advanced into the angioplasty balloon and allowed to coil (Fig. 6.2). At this point, the balloon was deflated. The guidewire was allowed to be drawn into the cannulation site as the balloon was retracted (Fig. 6.3). Once the guidewire was well within the vessel lumen, its further advancement was restrained, while the balloon was completely removed (Fig. 6.4).

With the guidewire in place, the remainder of the procedure directed toward the treatment of the juxta-anastomotic venous stenosis progressed in the usual manner.

G.A. Beathard, MD, PhD (✉)
Lifetime Vascular Access, University of Texas Medical Branch,
5135 Holly Terrace, Houston, TX 77056, USA
e-mail: gbeathard@msn.com

© Springer International Publishing AG 2017
A.S. Yevzlin et al. (eds.), *Dialysis Access Cases*,
DOI 10.1007/978-3-319-57500-1_6

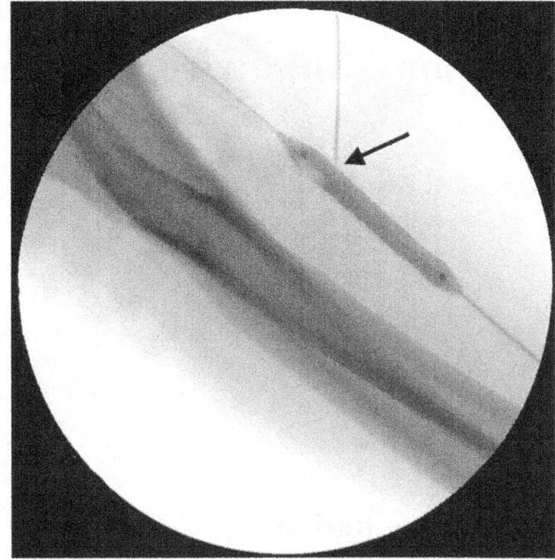

Fig. 6.1 Puncture of angioplasty balloon with micropuncture needle (*arrow*)

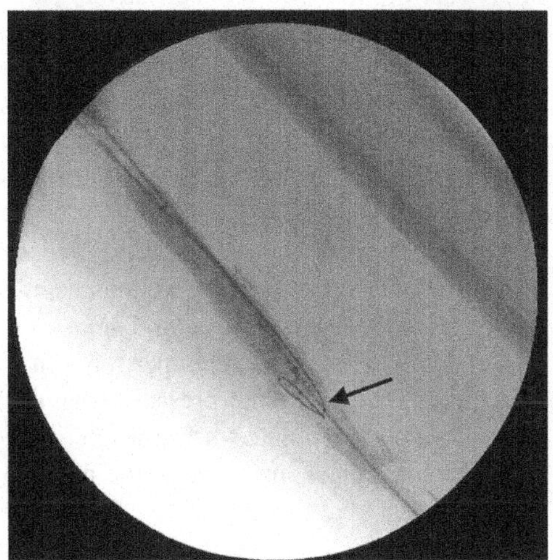

Fig. 6.2 Guidewire coiling (*arrow*) as it is advanced into angioplasty balloon

Discussion

There are instances in which cannulation of a target vessel is difficult either because a vessel is occluded or it cannot be adequately defined. In these instances, either an endovascular snare [1–3] or an inflated angioplasty balloon [4–7] has been placed in the target vessel lumen as a cannulation guide. In the case of the snare, not only is it used as a cannulation target, but it is also used as a carrier of the guidewire as it is retracted into the vascular lumen. The same is true for the angioplasty balloon; it can be utilized to retract the guidewire once it has become entangled within the perforated balloon.

Fig. 6.3 The deflated angioplasty balloon is being retracted carrying guidewire with it. *Arrow* indicates the most proximal balloon marker

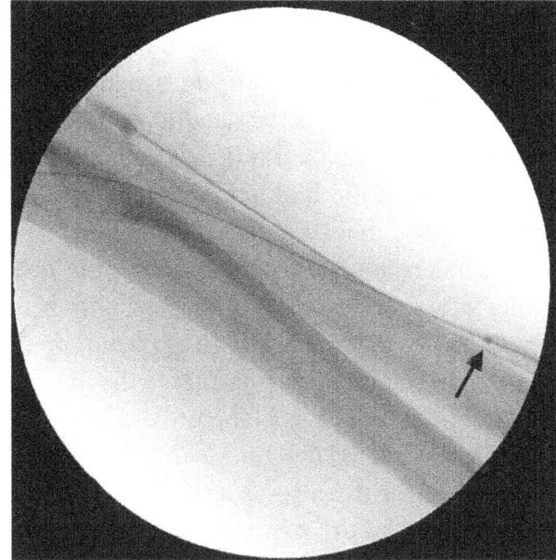

Fig. 6.4 Angioplasty balloon has been withdrawn, and the micropuncture guidewire can be seen within the vessel lumen with its the angioplasty actually extending beyond the initial cannulation site (*arrows*)

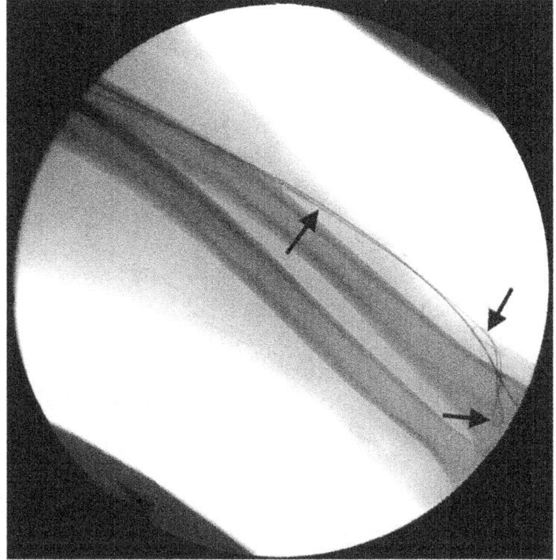

In this case, it was necessary to cannulate the downstream portion of the arteriovenous fistula in a retrograde direction in order to gain access to the juxta-anastomotic stenosis. However, the lack of development of the downstream portion of the AVF in addition to poor blood flow and pressure made cannulation of this segment very difficult. In such a situation, there are two alternatives – approach the lesion from an arterial cannulation or use balloon-guidewire entrapment to gain entry into the poorly developed segment.

Fortunately, even in cases in which the downstream segment is poorly developed, there is generally an area in the region of the anastomosis where antegrade cannulation can be easily accomplished. The problem with this site is that it is too close to the lesion to allow for the placement of a sheath. However, this site can be used to gain entry into the downstream segment using the balloon-guidewire entrapment maneuver with the angioplasty balloon serving as a target.

This procedure is accomplished by first cannulating the available juxta-anastomotic site with the needle pointing downstream. Once a guidewire has been advanced through the AVF into the draining veins, an angioplasty balloon is then inserted. Once the balloon has been advanced to the desired cannulation site and inflated to a point of complete effacement, it was used as a target for cannulation with a micropuncture needle. Once the needle entered the target, the 0.018-inch guidewire was advanced into the balloon where it became coiled and entangled. Then the deflated balloon with its entrapped guidewire was retracted back toward the original cannulation site. Once the micropuncture guidewire was well within the lumen of the AVF, it was then restrained as further retraction of the balloon continued. This extracted the guidewire from the balloon leaving it within the lumen of the AVF ready for the insertion of the micropuncture dilator. From this point forward, the treatment of the stenotic lesion progressed in the usual fashion.

References

1. Nazarian GK, Myers TV, Bjarnason H, et al. Applications of the Amplatz snare device during interventional radiologic procedures. AJR Am J Roentgenol. 1995;165:673.
2. Selby JB, Tegtmeyer CJ, Amodeo C, et al. Insertion of subclavian hemodialysis catheters in difficult cases: value of fluoroscopy and angiographic techniques. AJR Am J Roentgenol. 1989;152:641.
3. Hagen P, Yang JJ, Saibil EA. Use of an Amplatz goose neck snare as a target for collateral neck vein dialysis catheter placement. J Vasc Interv Radiol. 2001;12:493.
4. Horton MG, Mewissen MW, Rilling WS, et al. Hemodialysis catheter placement directly into occluded central vein segments: a technical note. J Vasc Interv Radiol. 1999;10:1059.
5. Farrell T, Lang EV, Barnhart W. Sharp recanalization of central venous occlusions. J Vasc Interv Radiol. 1999;10:149.
6. Gupta H, Murphy TP, Soares GM. Use of a puncture needle for recanalization of an occluded right subclavian vein. Cardiovasc Intervent Radiol. 1998;21:508.
7. Chen MC, Chang SC, Weng MJ, et al. Use of angioplasty balloon-assisted Seldinger technique for complicated small vessel catheterization. J Vasc Interv Radiol. 2006;17:2011.

Chapter 7
Angioplasty of Juxta-anastomotic Stenosis with Venous Rupture

Gerald A. Beathard

Case Presentation

The patient was a 72-year-old male dialysis patient with an 8-week goal radial-cephalic arteriovenous fistula (AVF). The patient was referred to the dialysis access facility for non-maturation of the AVF. Physical examination of the anastomosis of the AVF revealed that it was hyper-pulsatile with a systolic thrill only. The bruit was high pitched and systolic only. The cephalic vein above the anastomosis was not palpable. Doppler ultrasound examination revealed the presence of a juxta-anastomotic stenosis.

After the patient was prepped and draped, the AVF was cannulated at the mid-forearm level in a retrograde direction, and an angiogram was performed. This showed the presence of juxta-anastomotic venous stenosis (Fig. 7.1, arrow). A guidewire was passed down the AVF and across the arterial anastomosis. Using a 6 × 4 angioplasty balloon, the lesion was successfully dilated to full balloon effacement (Fig. 7.2). A post-procedure angiogram was performed by compressing the upper portion of the fistula and injecting radiocontrast retrograde. This showed good results initially (Fig. 7.3); however, a repeat injection was performed in the same manner resulting in extravasation of contrast (Fig. 7.4). Because of persistent problems with blood flow following balloon tapenade of the site, endovascular stent was placed. This gave a good result in the patient who was returned to the dialysis clinic.

G.A. Beathard, MD, PhD (✉)
Lifetime Vascular Access, University of Texas Medical Branch,
5135 Holly Terrace, Houston, TX 77056, USA
e-mail: gbeathard@msn.com

© Springer International Publishing AG 2017
A.S. Yevzlin et al. (eds.), *Dialysis Access Cases*,
DOI 10.1007/978-3-319-57500-1_7

Fig. 7.1 Initial angiogram showing juxta-anastomotic stenosis (*arrow*)

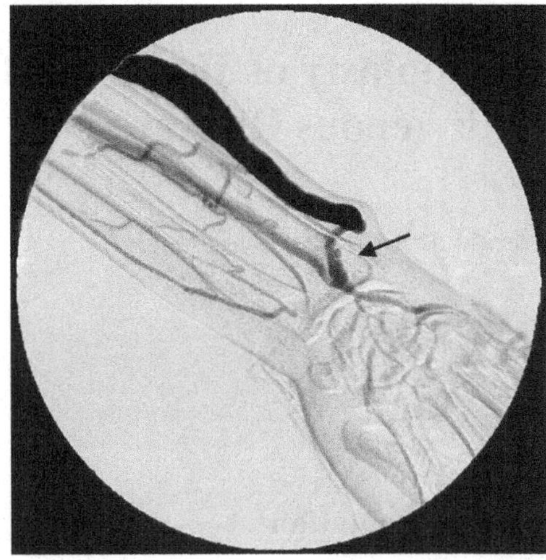

Fig. 7.2 Angioplasty of lesions showing total effacement of angioplasty balloon

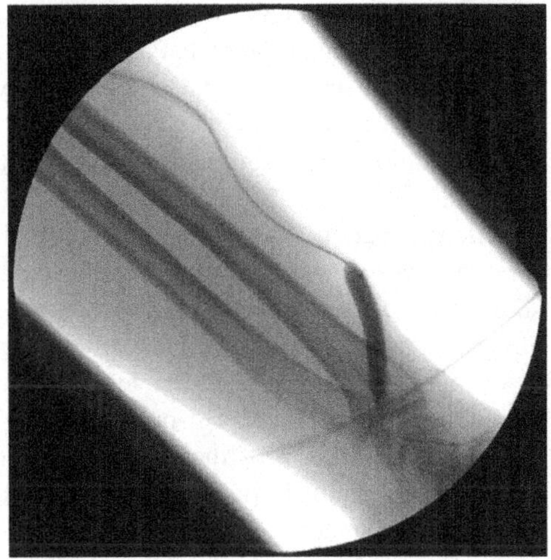

Discussion

The most common lesion resulting in failure of an AVF to mature is stenosis of the juxta-anastomotic segment of the AVF [1–5]. In order to gain access to the site of the lesion, a retrograde cannulation of the upper part of the AVF is commonly performed. Visualization of the area of pathology is generally obtained by retrograde injection of radiocontrast with manual occlusion of the upper AVF. Although this applies considerable pressure to the lower portion of the AVF in order to overcome the blood pressure, problems are rarely encountered. However, successful

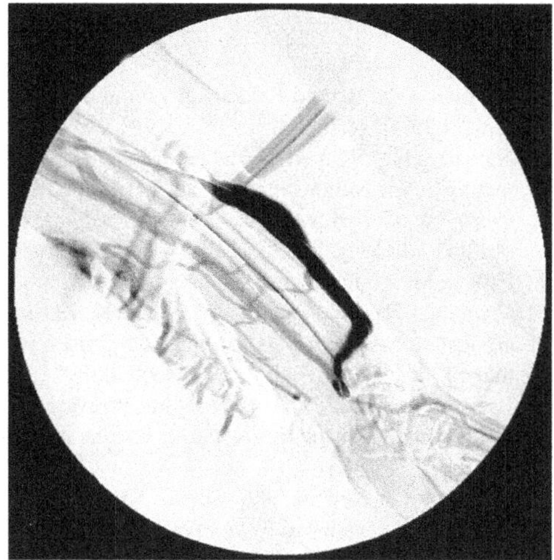

Fig. 7.3 Initial appearance of post-procedure angiogram showing good result of treatment

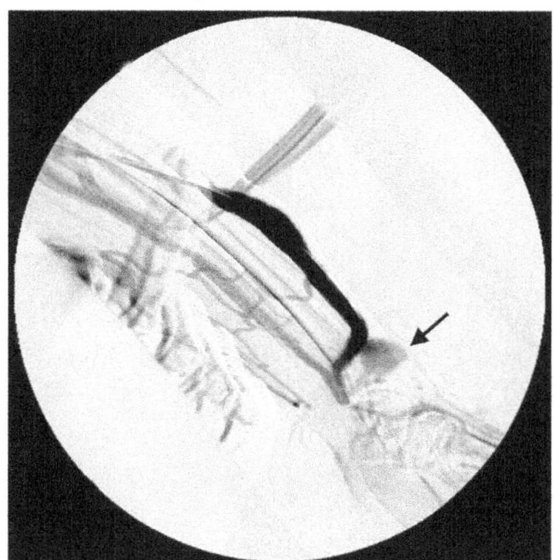

Fig. 7.4 Follow-up angiogram showing venous rupture with extravasation of radiocontrast (*arrow*)

angioplasty results in mechanical injury to the vessel wall [6, 7]. This weakens the vessel and makes it susceptible to rupture when subjected to excessive pressure which has occurred in this case.

In order to avoid this complication which can result in loss of the access, a post-angioplasty angiogram following the treatment of juxta-anastomotic stenosis should not be performed with a retrograde radiocontrast injection using manual occlusion of the fistula [8]. A better solution is to pass a vascular catheter across the arterial anastomosis and perform an antegrade angiogram through the catheter.

References

1. Beathard GA, Arnold P, Jackson J, et al. Aggressive treatment of early fistula failure. Kidney Int. 2003;64:1487.
2. Nassar GM, Nguyen B, Rhee E, Achkar K. Endovascular treatment of the "failing to mature"arteriovenous fistula. Clin J Am Soc Nephrol. 2006;1:275.
3. Clark TW, Cohen RA, Kwak A, et al. Salvage of nonmaturing native fistulas by using angioplasty. Radiology. 2007;242:286.
4. Falk A. Maintenance and salvage of arteriovenous fistulas. J Vasc Interv Radiol. 2006;17:807.
5. Manninen HI, Kaukanen E, Makinen K, Karhapaa P. Endovascular salvage of nonmaturing autogenous hemodialysis fistulas: comparison with endovascular therapy of failing mature fistulas. J Vasc Interv Radiol. 2008;19:870.
6. Higuchi T, Okuda N, Aoki K, et al. Intravascular ultrasound imaging before and after angioplasty for stenosis of arteriovenous fistulae in haemodialysis patients. Nephrol Dial Transplant. 2001;16:151.
7. Davidson CJ, Newman GE, Sheikh KH, et al. Mechanisms of angioplasty in hemodialysis fistula stenoses evaluated by intravascular ultrasound. Kidney Int. 1991;40:91.
8. Salman L, Asif A, Beathard GA. Retrograde angiography and the risk of arteriovenous fistula perforation. Semin Dial. 2009;22:698.

Chapter 8
Treatment of Accessory Vein with Embolization Coil

Gerald A. Beathard

Case Presentation

The patient was a 54-year-old male with a left arm radial-cephalic fistula (AVF). The AVF was 3 months old and had been used for the venous needle only at the dialysis. Cannulation in the upper part of the fistula was problematic, and the patient had experienced several infiltrations following failed attempts. Referral to the dialysis access center was for non-maturation. Physical examination revealed that the thrill and bruit at the anastomosis were low pitched and continuous. The lower part of the fistula was visible; the upper portion was not. Pulse augmentation was good. Examination of Doppler ultrasound revealed a large accessory cephalic vein originating in the midforearm. It was the same size as the cephalic vein.

After the patient was prepped and draped, the AVF was cannulated in the upper forearm in a retrograde direction. An angiogram was performed which revealed the presence of the large accessory cephalic vein (Fig. 8.1). There was no evidence of upstream venous stenosis. The accessory vein was selectively catheterized (Fig. 8.2). An 8 × 5 embolization coil was delivered into the proximal portion of the vein (Fig. 8.3). A post-procedure angiogram showed good coil placement and the desired effect on blood flow (Fig. 8.4).

G.A. Beathard, MD, PhD (✉)
Lifetime Vascular Access, University of Texas Medical Branch,
5135 Holly Terrace, Houston, TX 77056, USA
e-mail: gbeathard@msn.com

© Springer International Publishing AG 2017
A.S. Yevzlin et al. (eds.), *Dialysis Access Cases*,
DOI 10.1007/978-3-319-57500-1_8

Fig. 8.1 Pre-procedure
angiogram showing
appearance of accessory
vein (*arrow*)

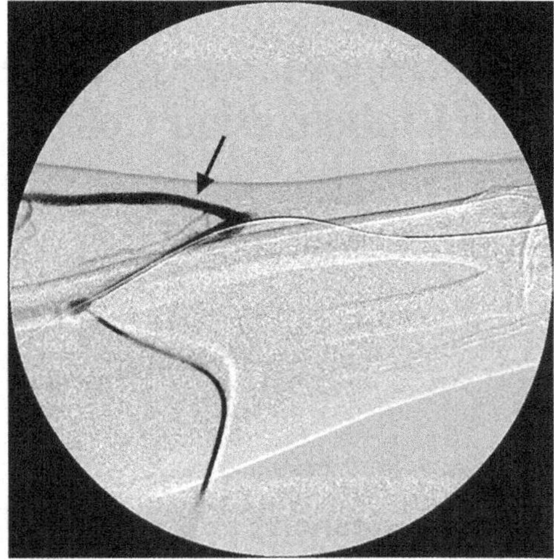

Fig. 8.2 Selective
catheterization of
accessory vein with the tip
of catheter in proximal
portion of vein

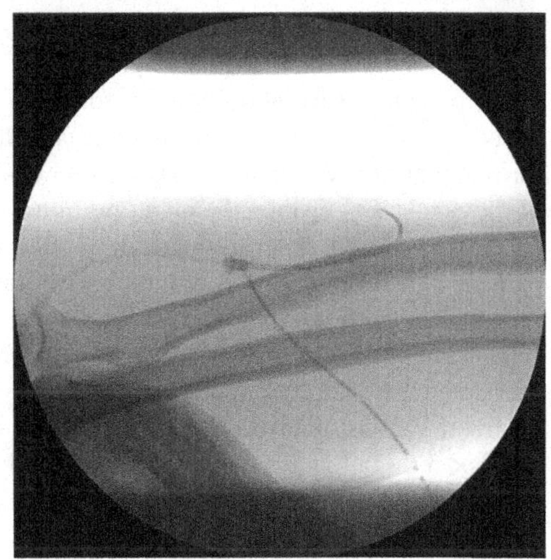

Discussion

The presence of a large accessory vein can adversely affect fistula (AVF) maturation
[1–7]. These are side branches of the forearm veins used for the construction of an
AVF. Accessory veins represent normal anatomy, and as such, they should not be
erroneously referred to as collateral veins which are pathological and related to
downstream stenosis. Unfortunately, the distinction is not always clear; stenosis can
make a normally present accessory vein larger and more prominent. The incidence
of accessory veins has been reported to be from 4% to 78% [2–7]. This marked variability probably represents differences in case selection. When all early failure cases

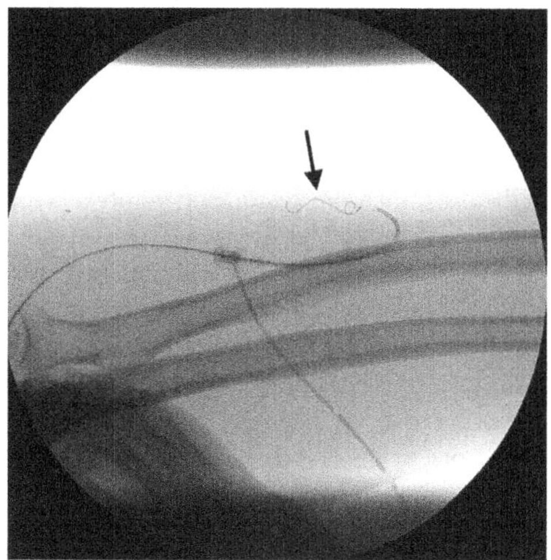

Fig. 8.3 Placement of embolization coil (*arrow*)

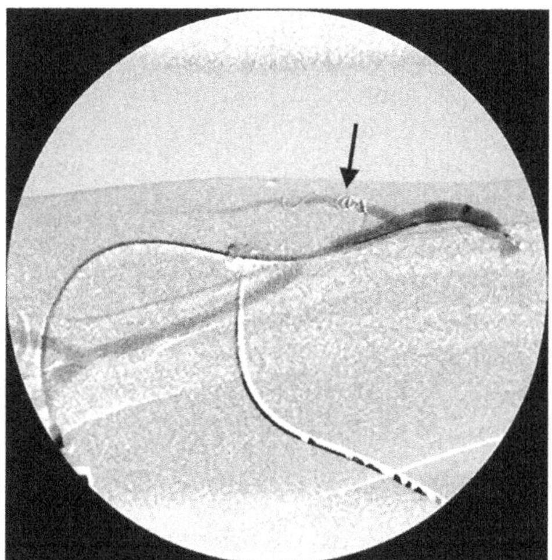

Fig. 8.4 Post-coil placement angiogram showing good coil placement (*arrow*) and very little flow in the accessory vein

are systematically evaluated, an incidence in the lower portion of this range can be anticipated. It should be mentioned that this is almost exclusively a problem with forearm AVFs. The upper arm superficial veins do not usually have any significant branches.

Preoperative assessment looking for accessory veins has been recommended [2]. It has been proposed that ligation of these vessels during initial AVF creation would potentially reduce non-maturation rates. Although this is frequently done; however, with increasing flow following the creation of an AVF, side branches that were initially judged to be without significance can become more prominent. Additionally, side branches that were not within the immediate surgical field may have been ignored.

The treatment of side branches is somewhat controversial. However, most investigators reporting series dealing with the non-maturing AVF have treated them [1–7]. When side branches are observed on the angiogram of an AVF, which has failed to mature, the first step is to determine if they are accessory or collateral veins. This is done by first determining if a downstream stenosis is present. If such is the case, then this lesion should always receive priority and be treated first to see if this will resolve the situation. The resistance with increased intraluminal pressure created by a stenotic lesion will not only cause the appearance of collateral veins but will also cause accessory veins to be larger than they actually will be when normal pressures are restored.

If there is no stenosis downstream from the side branch in question, or after the stenosis has been resolved if one is present, the next step is to decide the significance of the observed vessel. The mere presence of an accessory vein is not an adequate indication for its treatment. Small side branches are frequently encountered and are of no consequence. Additionally, there should be concern only if there is AVF dysfunction that can be attributed to the accessory vein. Such a vessel in an AVF that is functioning well may be an advantage, offering an additional area for cannulation and helping to maintain flow in the case of a thrombosis. Significance in the dysfunctional AVF should be judged by three criteria – size, flow, and changes that occur with manual occlusion. Unfortunately, there have been no clinical trials to provide guidance in this matter. The judgments made are somewhat subjective but should be based upon sound reasoning.

Changes that occur with manual compression of the side branch are often best evaluated by comparing the sound over the AVF above the accessory vein with and without compression of the vessel in question using a vascular Doppler. Flow can be determined using Doppler ultrasound with and without downstream compression. It can also be judged subjectively by comparing flow in the accessory and the main AVF side by side with a small bolus of radiocontrast. While this is subjective, it does add weight to a decision to obliterate when other factors are also supportive. One study [2] found that preoperatively detected accessory veins with a diameter >70% of the cephalic vein diameter had a sensitivity, specificity, positive predictive value, and negative predictive value of 80%, 100%, 100%, and 91%, respectively, for prediction of radial-cephalic non-maturation. However, this was based upon a rather small cohort of patients (4 of 10 non-maturing AVFs). We have used 25% of the diameter of the AVF as an indicator of possible significance in the face of a non-maturing AVF.

It seems reasonable that Poiseuille's law would be a good guide in determining the significance of an accessory vein. This would suggest that if the ratio between the diameter of the side branch and the main trunk of the AVF was 1:4, the flow would be 1:256, 1:3 would be 1:81, and 1:2 would be 1:16. However, this is not the case; Poiseuille's law does not apply here. The assumptions of the equation are that the flow is laminar and that the flow is through a rigid circular cross section that is substantially longer than its diameter. None of these assumptions hold for the accessory vein. Additionally, an accessory vein often leads to a field of veins; this fact also disqualifies Poiseuille's law. Basically, one is left with the need for making a subjective judgment based upon size, flow, and changes that occur with manual occlusion in an AVF that is dysfunctional.

There are three techniques that may be used for accessory vein obliteration – percutaneous ligation, surgical ligation, and the use of an embolization coil. The best technique to use depends upon the individual case situation.

References

1. Faiyaz R, Abreo K, Zaman F, et al. Salvage of poorly developed arteriovenous fistulae with percutaneous ligation of accessory veins. Am J Kidney Dis. 2002;39:824.
2. Planken RN, Duijm LE, Kessels AG, et al. Accessory veins and radial-cephalic arteriovenous fistula non-maturation: a prospective analysis using contrast-enhanced magnetic resonance angiography. J Vasc Access. 2007;8:281.
3. Beathard GA, Arnold P, Jackson J, et al. Aggressive treatment of early fistula failure. Kidney Int. 2003;64:1487.
4. Beathard GA, Settle SM, Shields MW. Salvage of the nonfunctioning arteriovenous fistula. Am J Kidney Dis. 1999;33:910.
5. Clark TW, Cohen RA, Kwak A, et al. Salvage of nonmaturing native fistulas by using a8.ngioplasty. Radiology. 2007;242:286.
6. Falk A. Maintenance and salvage of arteriovenous fistulas. J Vasc Interv Radiol. 2006;17:807.
7. Nassar GM, Nguyen B, Rhee E, Achkar K. Endovascular treatment of the "failing to mature" arteriovenous fistula. Clin J Am Soc Nephrol. 2006;1:275.

Chapter 9
Using a Snare to Extract Misplaced Coil

Gerald A. Beathard

Case Presentation

The patient is a 43-year-old male dialysis patient with a non-maturing left brachial-cephalic arteriovenous fistula (AVF). Physical examination of the AVF revealed good pulse augmentation, and the fistula collapsed with arm elevation. The lower portion of the fistula was easily palpable and had been used for the venous needle at dialysis. The nurses were unable to cannulate the upper part of the AVF. The accessory vein was evident by physical examination. Doppler ultrasound examination showed the presence of a large vein coming off of the cephalic vein at the level of the mid-upper arm. This vein was 4 mm in diameter.

After the patient was prepped and draped, the AVF was cannulated in an antegrade direction and the sheath was put in place. An angiogram was performed which showed the presence of the large accessory vein. This was selectively catheterized, and a 6 × 8 embolization coil was placed in the proximal segment of the vein. A decision was made to insert a second coil. Once this coil was released, it was apparent that the second coil was extending into the lumen of the AVF (Fig. 9.1, small arrow). An Amplatz gooseneck snare was deployed in order to extract the misplaced coil. The first step in the procedure involved the introduction of a guidewire which was advanced beyond the position of the coil followed by a vascular catheter through which the coil was passed and deployed (Fig. 9.1, large arrow). The deployed snare was then pulled back until it engaged the coil (Fig. 9.2, large arrow). At this point, the vascular catheter through which the snare had been passed was advanced while holding the shaft of the snare stationary. This had the effect of

G.A. Beathard, MD, PhD (✉)
Lifetime Vascular Access, University of Texas Medical Branch,
5135 Holly Terrace, Houston, TX 77056, USA
e-mail: gbeathard@msn.com

© Springer International Publishing AG 2017
A.S. Yevzlin et al. (eds.), *Dialysis Access Cases*,
DOI 10.1007/978-3-319-57500-1_9

Fig. 9.1 Gooseneck snare (*arrow*) has been advanced beyond problem coil and deployed

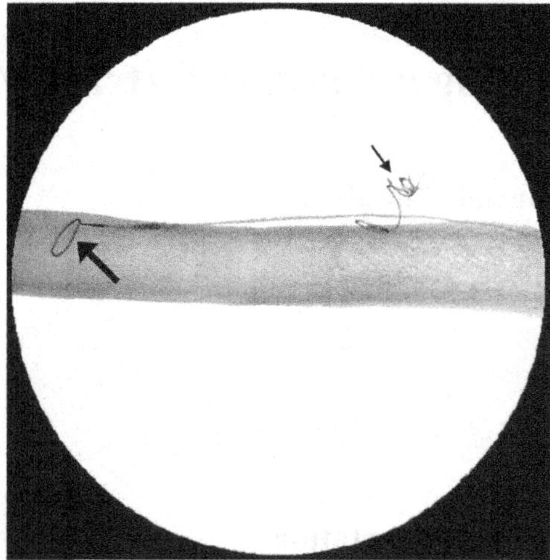

Fig. 9.2 Snare has been retracted and has caught the coil and the catheter has been advanced to compress the snare (*arrow*)

closing the snare tightly onto the coil (Fig. 9.3, large arrow). The snare with the entrapped coil (Fig. 9.4, large arrow) was then extracted.

This case involving the obliteration of an accessory vein with a coil was then completed in the usual fashion without further problems.

Discussion

Since its first description, the percutaneous retrieval of intravascular foreign bodies has become a frequently used technique. Although other devices can be used, the endovascular snare has become the standard. While a variety of snares are available,

Fig. 9.3 Snare-catheter ensemble is being retracted carrying the coil with it (*arrow*)

Fig. 9.4 Coil is in process of being extracted (*arrow*)

the Amplatz gooseneck snare (Microvena, St. Paul, Minnesota) and the EnSnare® (Merit Medical Systems, Malvern, Pennsylvania) are the two most frequently used in association with dialysis access procedures. One of these devices should be in the armamentarium of every interventionalist dealing with these types of problems.

The gooseneck snare [1] has been in use for a longer period; however, with its three-loop design, the EnSnare is somewhat more easily used. Except for the number of loops, the construction and use of these two types of snares are very similar. Both snares are constructed of nitinol which provides a degree of kink resistance, strength, and torque control to the device. Because nitinol is poorly radiopaque, the gooseneck snare contains gold-tungsten coils, and the EnSnare contains platinum coils to enhance visualization. Both snares can be obtained in a variety of loop sizes.

The snare size used should be chosen based upon the size of the vessel involved. Generally, the 10 and 15 mm sizes appear to be the best suited for dialysis access use.

The device comes in a kit which includes the snare mounted onto a shaft used for manipulation, a catheter which is used for the insertion of device, an introducer which is used to compress the snare for insertion into the delivery catheter, and a torquing device which is used to manipulate the snare once it has been inserted. The catheter that comes with the kit varies in size according to the dimensions of the snare which is included. However, any catheter of appropriate size can be used for the introduction of the snare.

In order to accomplish the procedure using the snare, a vascular catheter for its introduction must first be put in place over a guidewire which has been passed beyond the target site (in this case, the misplaced coil). Once this catheter is in place beyond the target site, the snare is introduced into the vascular catheter by first compressing it using the introducer that comes with the kit. Using the shaft of the snare, it is advanced beyond the tip of the vascular catheter where it deploys. At this point, the snare edges will be pressed against the wall of the vessel, providing that the choice of size has been correctly made. Leaving the snare stationary, the vascular catheter should be withdrawn so that its tip lies just peripheral to the location of the coil. The snare should then be slowly retracted in a deployed configuration by withdrawing the shaft to which it is connected back into the snare catheter which is being held steady. This process is continued until the snare catches (snares) the coil. This may require some manipulation using the torquing device. If the snare is retracted beyond the coil without catching it, the process should be started again. Once the snare has caught the coil, its position must be maintained with one hand, while the vascular catheter is advanced with the other hand in order to close the loop of the snare. The coil will be entrapped by the snare as it is compressed by the advancing catheter tip. Attempting to close the snare by pulling it into the catheter will displace the loop of the snare from its position around the coil. Once the coil has been captured, forward pressure should be maintained on the catheter in order to keep the snare tightly compressed as the entire ensemble is slowly withdrawn and the coil has been extracted from the patient.

Reference

1. Yedlicka JW Jr, Carlson JE, Hunter DW, et al. Nitinol gooseneck snare for removal of foreign bodies: experimental study and clinical evaluation. Radiology. 1991;178:691.

Chapter 10
Basilic Vein Transposition

Kathleen M. Lamb and Paul J. Foley

Abbreviations

AVF   Arteriovenous fistula
BVT   Basilic vein transposition

Case Presentation 1

This patient was a 59-year-old left-handed male who had end-stage renal disease requiring hemodialysis. He was in need of more permanent dialysis access.

On physical examination, the patient was found to have no prior upper extremity scars or previously placed arteriovenous fistula (AVF). He had palpable radial and ulnar pulses bilaterally and superficial venous engorgement noted when a blood pressure cuff was elevated on his arms. Vein mapping was performed, demonstrating good caliber cephalic and basilic vein (Fig. 10.1). Based on his preoperative vein mapping studies and physical exam, it was recommended that he undergo a right upper extremity arteriovenous fistula creation.

The patient's right arm was prepped and draped; a transverse incision was made over the proximal forearm. Sharp dissection of the cephalic vein was carried out, but this vein was ultimately found to be sclerotic with multiple branches, none of which

K.M. Lamb, MD
Hospital of the University of Pennsylvania, Department of Vascular and Endovascular Surgery, Philadelphia, PA, USA

P.J. Foley, MD (✉)
Department of Vascular and Endovascular Surgery, Hospital of the University of Pennsylvania, 4 Silverstein, 3400 Spruce St, Philadelphia, PA 19104, USA
e-mail: Paul.foley@uphs.upenn.edu

© Springer International Publishing AG 2017
A.S. Yevzlin et al. (eds.), *Dialysis Access Cases*,
DOI 10.1007/978-3-319-57500-1_10

	Thrombus		Doppler Pulsatility	
	Right	Left	Right	Left
Internal Jugular Vein	No	No	Normal	Normal
Innominate Vein	No	No	Normal	Normal
Subclavian Vein	No	No	Normal	Normal
Axillary Vein	No	No	Normal	Normal
Superior Vena Cave				

| | Right | | Left | |
	Diameter (cm)	Depth (cm)	Diameter (cm)	Depth (cm)
Cephalic Vein				
Shoulder	0.26	1.16		
Mid Upper Arm	0.24	0.18	0.11	0.74
Antecubital	0.27	0.21		
Mid Forearm	0.20	0.17		
Wrist	0.33	0.25		
Basilic Vein				
Axillary	0.89	0.62	0.93	0.55
Mid Upper Arm	0.56	0.79	0.79	0.79
Antecubital Fossa	0.54	0.59	0.82	0.50
Mid Forearm			0.52	0.49
Wrist			0.51	0.25

Fig. 10.1 Preoperative vein mapping

were of an adequate caliber for an AVF. Through the same incision, a large basilic vein was identified that would be suitable for fistula creation. The decision was made to proceed with a brachiobasilic anastomosis as the first stage of an eventual basilic vein transposition (BVT). This basilic branch was dissected free and the brachial artery was identified. Proximal and distal arterial control was obtained with vessel loops, and the basilic vein branch was divided and flushed with heparinized saline. An arteriotomy was made and the vein was spatulated for an end-side anastomosis. At the conclusion of the anastomosis, the vessel loops were released, and a strong palpable thrill could be felt just proximal to the incision, and a palpable radial pulse was felt at the wrist.

Two months later, this patient was taken back to the operating room to superficialize the basilic vein. The basilic vein felt to have dilated sufficiently (approximately 6 mm) based on physical exam. A longitudinal incision was made along the length of the basilic vein in the upper arm (Fig. 10.2). A subcutaneous pocket was dissected out and the deeper tissues were closed, elevating the basilic vein (Fig. 10.3). The vein was tucked in the superficial skin pockets and the incision closed above. On follow-up, the mature BVT AVF was suitable for use for hemodialysis.

Discussion

Basilic vein transpositions are commonly used for upper extremity hemodialysis access, which can be performed as either a single or two-staged procedure. The decision regarding when to perform single or two-staged procedures is multifactorial.

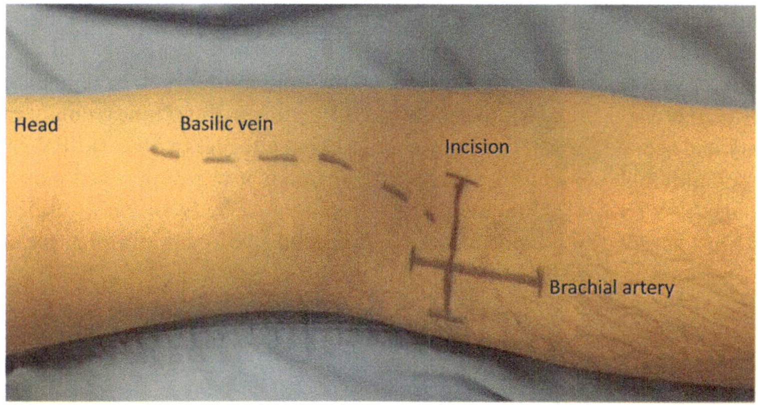

Fig. 10.2 Marking of planned incision

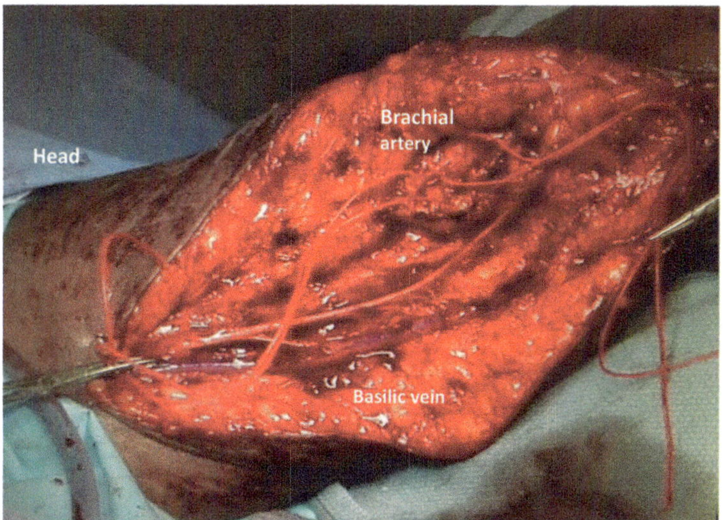

Fig. 10.3 Exposure of brachial artery and basilic vein

Transposing the basilic vein allows access to a longer length of fistula in a more superficial location. This makes the fistula easier to access for hemodialysis, especially given the rise of obesity nationwide [1]. Transposition also enables ligation of vein branches to decrease potential steal of blood flow within the fistula [1]. Two-staged procedures have the benefit of allowing time for vein wall arterialization, increasing wall thickness and diameter. This is useful if the basilic vein is initially small in caliber (<4 mm) [2]. Two-staged procedures may avoid large incisions in patients who may not mature the AVF during the first stage of the operation, especially as many of these patients have comorbidities associated with poor wound healing. Less favorable factors related to a two-staged procedure are longer times until AVF is usable for hemodialysis, potentially increased morbidity and increased cost given the need for a second operation, and possibly hospitalization [2]. However, there appears to be no significant difference in the primary failure rates or primary and secondary patency rates comparing single and two-staged BVT operations [2], although some studies demonstrate trends toward improved patency for two-staged BVT [3, 4]. Ultimately, the decision must be made on a case-by-case basis. Most surgeons recommend waiting 4–6 weeks before performing the second stage operating in two-staged BVT [2].

Case Presentation 2

A 53-year-old male with renal failure requiring hemodialysis presented for placement of permanent dialysis access. Preoperative vein mapping demonstrated adequate basilic forearm vein, and it was recommended that he undergo creation of a left upper extremity AVF (Fig. 10.4). Preoperatively, his basilic vein was marked with ultrasound from the elbow to the wrist. An upper extremity arterial ultrasound was also performed, demonstrating radial artery occlusion in the distal forearm. The decision was made to proceed with a forearm basilic vein transposition, utilizing a gentle forearm loop configuration with a basilic vein to brachial artery anastomosis.

	Thrombus		Doppler Pulsatility	
	Right	Left	Right	Left
Internal Jugular Vein	No	Yes	Normal	Normal
Innominate Vein	No	No	Normal	Normal
Subclavian Vein	No	No	Normal	Normal
Axillary Vein	No	No	Normal	Normal
Superior Vena Cave				

	Right		Left	
	Diameter (cm)	Depth (cm)	Diameter (cm)	Depth (cm)
Cephalic Vein				
Shoulder			0.44	1.02
Mid Upper Arm			0.24	0.52
Antecubital	0.23	1.75	0.23	0.86
Mid Forearm				
Wrist				
Basilic Vein				
Axillary	0.79	0.94	0.58	0.86
Mid Upper Arm	0.31	0.93	0.42	0.90
Antecubital Fossa	0.56	0.85	0.47	0.61
Mid Forearm	0.32	0.17	0.38	0.22
Wrist	0.24	0.28	0.37	0.27

Fig. 10.4 Preoperative vein mapping

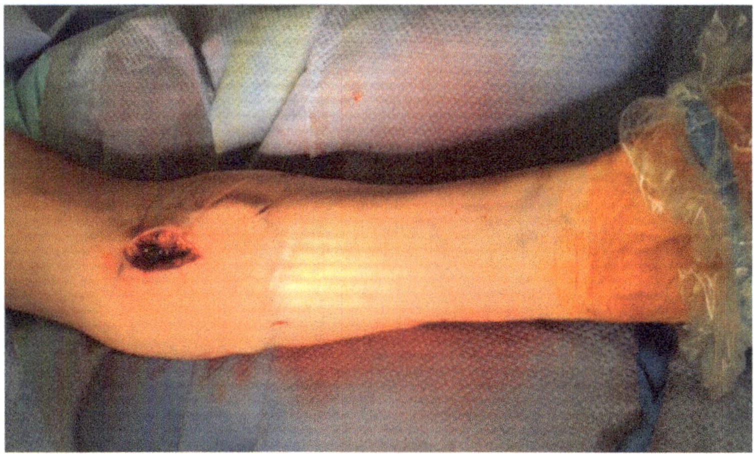

Fig. 10.5 Brachial artery exposure

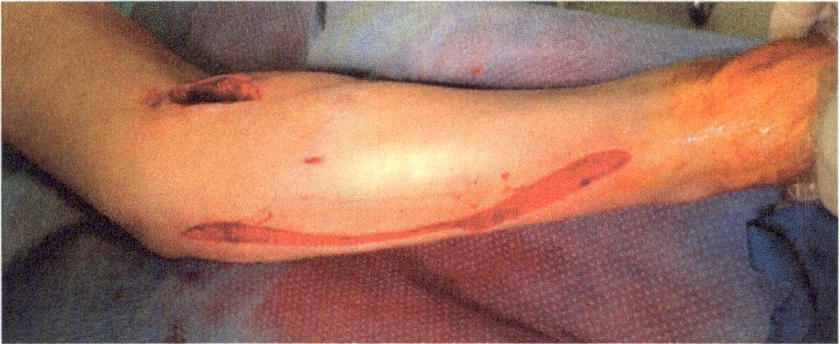

Fig. 10.6 Brachial artery exposure with basilic artery harvest in the forearm

A vertical incision was made along the proximal forearm over the distal brachial artery, which was sharply dissected; proximal and distal control was obtained (Fig. 10.5). A second incision was made along the basilic vein from the elbow to wrist. The basilic vein was dissected out along the course of the incision, and all branches were ligated and divided (Fig. 10.6). The basilic vein was then divided at the wrist and flushed with heparinized saline. A loop was created and tunneled to extend over the forearm. A counter-incision was made to facilitate loop formation, and care was taken to ensure the basilic vein was tunneled in the appropriate orientation without twisting or kinking (Fig. 10.7). An end-to-side anastomosis was then created from the basilic vein to the brachial artery. After completing the anastomosis, a strong thrill was palpable along the fistula.

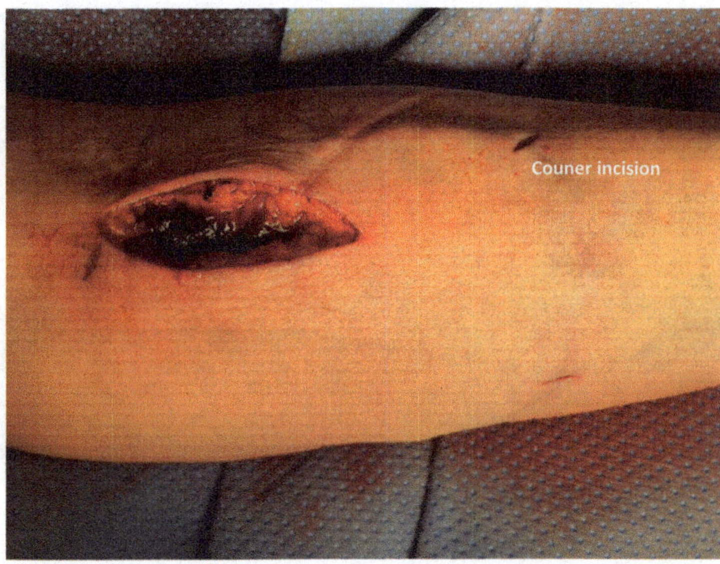

Fig. 10.7 Basilic vein transposition after graft tunneling. Note: relaxing incisions along the graft for ease of tunneling and to prevent graft kinking or twisting

Discussion

Hemodialysis access can often be challenging and is dependent on a variety of patient factors. Preoperative physical examination of the extremities with and without a tourniquet is helpful for guiding decision-making regarding optimal arteriovenous fistula placement. Important information can also be obtained through a detailed history of any prior hemodialysis access surgery or other relevant procedures performed on the extremities such as peripherally inserted central catheters, permanent pacemaker, repeated venipuncture, and arterial lines. Preoperative vein mapping yields important information about the upper extremity veins with respect to diameter, patency, and depth from the skin. Mendes et al. [5] demonstrated a 76% fistula maturation rate if vein diameter is greater than 2 mm. Many surgeons, however, consider 3 mm to be the cutoff for suitable vein diameter. Additionally, artery size greater than 1.5–2 mm is ideal to achieve fistula maturation. This is typically only a concern when the radial artery is to be used, as brachial artery diameters are predominately of sufficient size. Small vessel caliber may reflect arterial occlusive disease and an inability of the vessel to distend with increased fistula flows, leading to fistula failure [6, 7].

Ultrasound can be used in the operating room prior to incision to confirm findings on preoperative ultrasound and to mark the vein and artery intended for use for ease of dissection. It is imperative, however, to evaluate the size and caliber of the vessels for use, which may ultimately change the preoperative plan for fistula placement. In this instance, the radial artery was unsuitable for use along the course of the forearm, and the brachial artery was selected instead. Additionally, the basilic vein in the forearm was of an appropriate size and was used in a nontraditional way as a forearm loop to the brachial artery.

Case Presentation 3

This patient is a 62-year-old right-handed man who presented as an outpatient with chronic renal insufficiency with worsening renal function. His nephrologist referred him to a clinic to evaluate for arteriovenous fistula placement pending the need for hemodialysis. He underwent vein mapping and was found to have suitable basilic vein bilaterally. On evaluation in the office, he had easily palpable left brachial and radial pulses.

Intraoperatively, a small, longitudinal incision was made over the brachial pulse, just above the left antecubital fossa; then proximal and distal control was obtained. Through this incision, the basilic vein was identified, traced, and dissected into the upper left arm. This incision was continued toward the axilla until a sufficient length of basilic vein was exposed to transpose (Fig. 10.8). Vein branches were ligated and divided until the basilic vein was free along its entire length, at which point the vein was divided at the antecubital crease (Fig. 10.9). The vein was distended with heparinized saline and marked for proper orientation, and the branch points were closely inspected. Using a tunneling device, a subcutaneous tunnel was created lateral to the

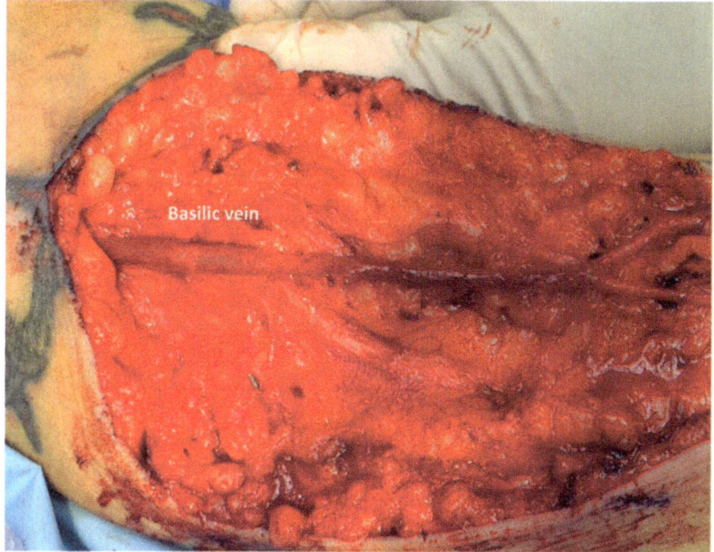

Fig. 10.8 Basilic vein dissected in situ

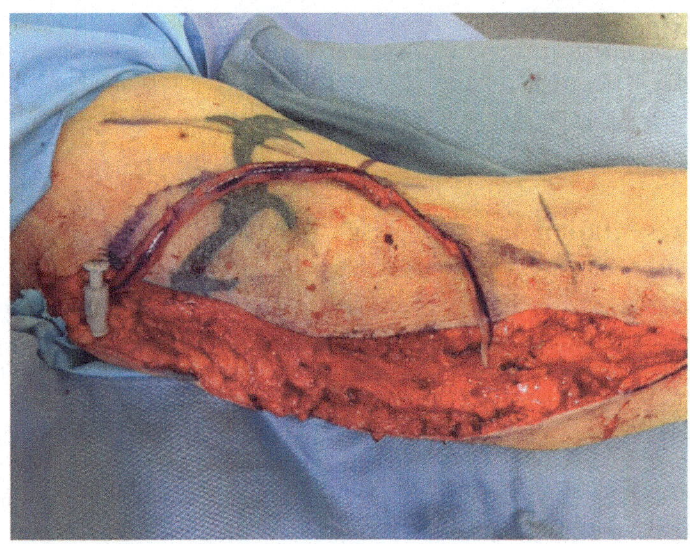

Fig. 10.9 Basilic vein ligated distally with layout of planned transposition

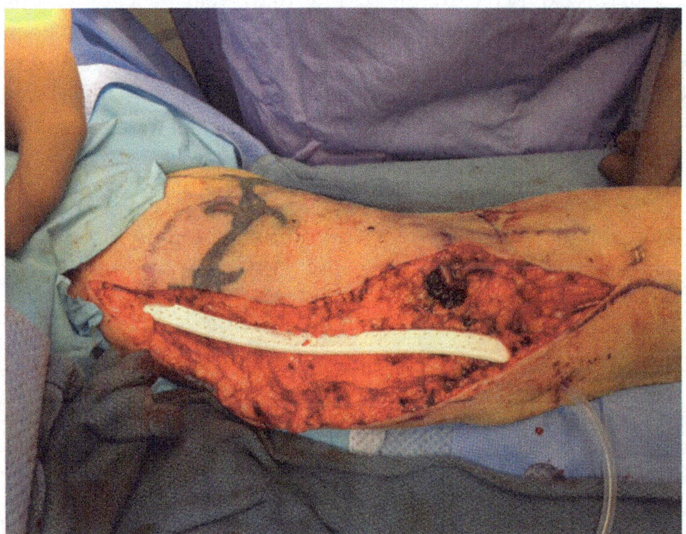

Fig. 10.10 Fistula after the basilic vein is transposed. Jackson-Pratt drain placed in dissection bed

incision, and the vein was pulled through the subcutaneous tunnel. The anastomosis was created between the ends of basilic vein to the side of the brachial artery using a running 6–0 Prolene suture. A strong thrill was felt in the fistula upon completion of the anastomosis, and there was a palpable radial pulse at the wrist. Hemostasis was achieved, and the wound was closed in multiple layers over a Jackson-Pratt drain (Fig. 10.10).

Discussion

The basilic and cephalic veins constitute the major superficial venous system draining the upper extremity. At the antecubital fossa, these veins usually branch to form the median cubital vein, while each additionally continues into the forearm, providing venous drainage of the hand. Both veins are routinely used for arteriovenous fistula creation, but the basilic vein is often a reasonable choice, as it is less traumatized by repeated venipuncture, given its deeper location in the upper extremity. Given the depth of the vein, it must be transposed or superficialized in order to provide adequate length to be accessed for hemodialysis.

Since the initial recommendation of the National Kidney Foundation's Kidney Disease Outcomes Quality Initiative (KDOQI) to have an arteriovenous fistula target prevalence of 40%, which has since been increased to 65%, arteriovenous fistula creation has been on the forefront of discussion among vascular surgeons. Where to place access and what access options exist if access fails are consuming questions. Most surgeons, however, agree that access should be obtained in more peripheral locations first and then progress up to the arm and then to other sites, such as the legs. Radiocephalic fistulas should therefore be attempted first, followed by brachiocephalic fistulas. Brachiobasilic fistulas, or basilic vein transpositions, although not typically a first choice for access, have advantages in that the veins are often less traumatized by repeated venipuncture, are often of a larger initial vein diameter, carry less risk of infection compared to prosthetic grafts, and have better primary and secondary patency rates than prosthetic grafts [8, 9]. Thrombosis has been cited as the most common complication leading to the inability of the fistula to mature, while central vein stenosis was the most common cause of fistula failure [8, 10].

Pearls

- Basilic vein transpositions can be performed as a single or a two-staged procedure with similar long-term patency rates.
- Benefits of basilic vein transposition hemodialysis access include decreased infection risk and improved patency rates compared to prosthetic grafts.
- Basilic vein is a good choice of conduit given thicker vein with wider diameter. The upper arm basilic vein will typically not have been traumatized by repeated venipuncture given the deeper location compared to the cephalic vein.

References

1. Korepta LM, Watson JJ, Elder EA, Davis AT, Monsour MA, Chambers CM, Cuff RF, Wong PY. Outcomes for forearm and upper arm arteriovenous fistula creation with the transposition technique. J Vasc Surg. 2016;63:764–71.
2. Cooper J, Power A, DeRose G, Forbes T, Dubois L. Similar failure and patency rates when comparing one- and two-stage basilic vein trans-positions. J Vasc Surg. 2015;61:809–16.

3. Vrakas G, Defiqueiredo F, Turner S, Jones C, Taylor J, Calder F. A comparison of the outcomes of one-stage and two-stage brachio-basilic arteriovenous fistulas. J Vasc Surg. 2013;58:1300–4.
4. Syed FA, Smolock CJ, Duran C, Anaya-Ayala JE, Naoum JJ, Hyunh EK, et al. Comparison of outcomes of one-stage basilic vein transpositions and two-stage basilic vein transpositions. Ann Vasc Surg. 2012;26:852–7.
5. Mendes RR, Farber MA, Marston WA, Dinwiddle LC, Keagy BA, Burnham SJ. Prediction of wrist arteriovenous fistula maturation with preoperative vein mapping with ultrasonography. J Vasc Surg. 2002;36:460–2.
6. Parmar J, Aslam M, Standfield N. Pre-operative radial arterial diameter predicts early failure of arteriovenous fistula (AVF) for haemodialysis. Eur J Vasc Endovasc Surg. 2007;33:113–5.
7. Huber TS. Hemodialysis access: general consideration. In: Rutherford's vascular surgery. 8th ed. Philadelphia: Elsevier. C; 2014.
8. Quinn B, Cull DL, Carsten CG. Hemodialysis access: placement and management of complications. In: Comprehensive vascular and endovascular surgery. 2nd ed. Philadelphia: Elsevier. C; 2009.
9. Tan TW, Farber A. Brachial-basilic autogenous access. Semin Vasc Surg. 2011;24:63–71.
10. Glass C, Porter J, Singh M, Gillespie D, Young K. Illig K. Ann Vasc Surg. 2010;24(1):85–91.

Chapter 11
Balloon-Assisted Banding for High-Output Heart Failure

Gerald A. Beathard

Case Presentation

The patient was a 79-year-old male who had been on dialysis for 4 years. Over the past 6 months, he has developed progressively more severe congestive heart failure manifesting as dyspnea on exertion and orthopnea. The patient's cardiac problems have been managed by a cardiologist who felt that he was receiving maximum medical management. His blood pressure was not elevated; he was not anemic and was not felt to be above his dry weight. The patient was referred to the dialysis access center because his nephrologist and cardiologist felt that he had a high blood flow fistula which was contributing to his cardiac failure.

Upon examination, the patient was found to have a left upper arm brachial-cephalic fistula which was large and dilated. Pulse augmentation was marked, the fistula was not hyper-pulsatile, and the thrill and bruit that were present over the fistula were very strong with both systolic and diastolic components. Ultrasound evaluation of access blood flow revealed a value of 3476 mL/min (Fig. 11.1). A decision was made to perform a balloon-assisted banding of the fistula to reduce the blood flow.

After the patient was prepped and draped, the fistula was cannulated in a retrograde direction, and a guidewire was advanced across the arterial anastomosis and up to the proximal portion of the brachial artery. An angiogram was performed to identify the anatomy of the fistula (Fig. 11.2). An incision was made over the fistula just proximal to the anastomosis, and a ligature was positioned around the fistula at that point. A 4 × 4 angioplasty balloon was then positioned in this area, inflated to maximum pressure (Fig. 11.3), and the ligature was tightened around the fistula to restrict it to the size of the balloon's diameter (Fig. 11.4). Ultrasound evaluation of

G.A. Beathard, MD, PhD (✉)
Lifetime Vascular Access, University of Texas Medical Branch,
5135 Holly Terrace, Houston, TX 77056, USA
e-mail: gbeathard@msn.com

© Springer International Publishing AG 2017
A.S. Yevzlin et al. (eds.), *Dialysis Access Cases*,
DOI 10.1007/978-3-319-57500-1_11

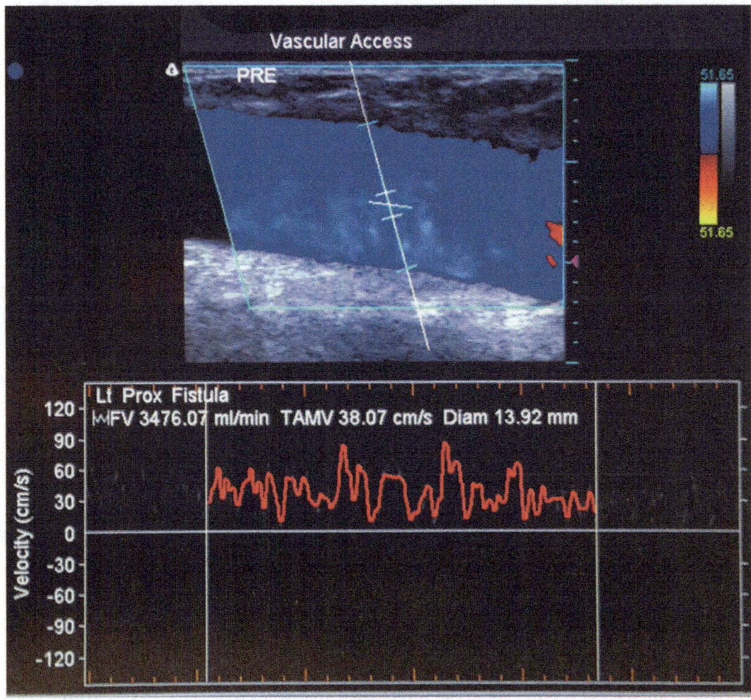

Fig. 11.1 Ultrasound determined access blood flow of 3476.07 mL/min

Fig. 11.2 Angiogram of
brachial-cephalic fistula

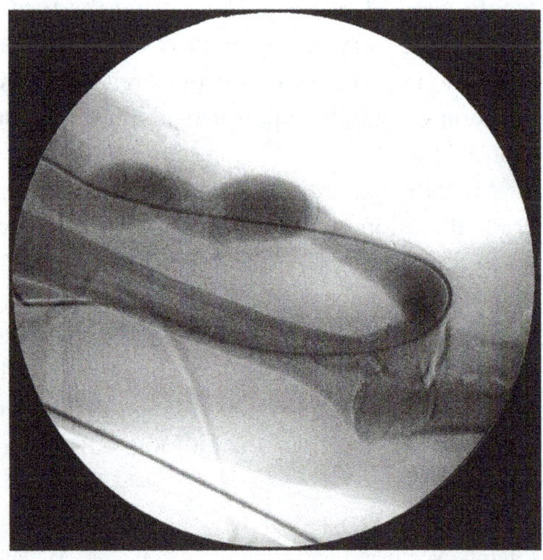

Fig. 11.3 Angioplasty balloon (4 × 4) in place. *A* angioplasty balloon

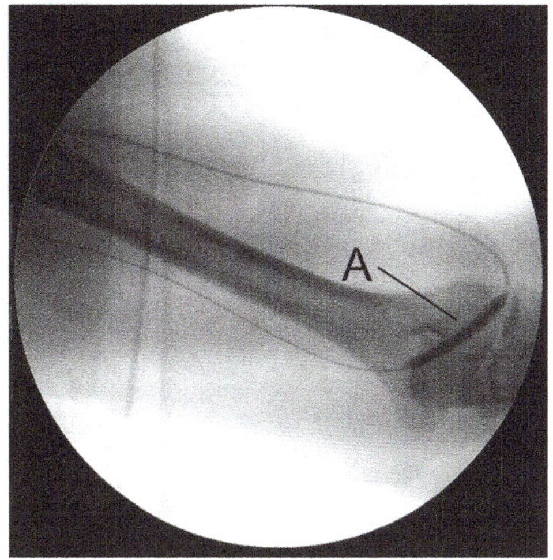

Fig. 11.4 Ligature around fistula

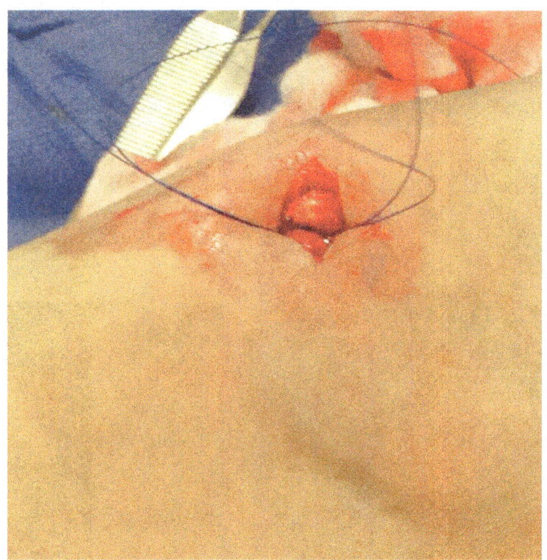

access blood flow was repeated and revealed a value of 995.5 mL/min (Fig. 11.5). The suture was then tied and the balloon was deflated and retracted. An angiogram was performed to evaluate the results of the banding procedure (Fig. 11.6). The incision was then closed.

A follow-up visit with the cardiologist 2 weeks after the banding procedure revealed that the patient's symptoms of congestive heart failure had resolved. The patient continued to dialyze successfully with his fistula.

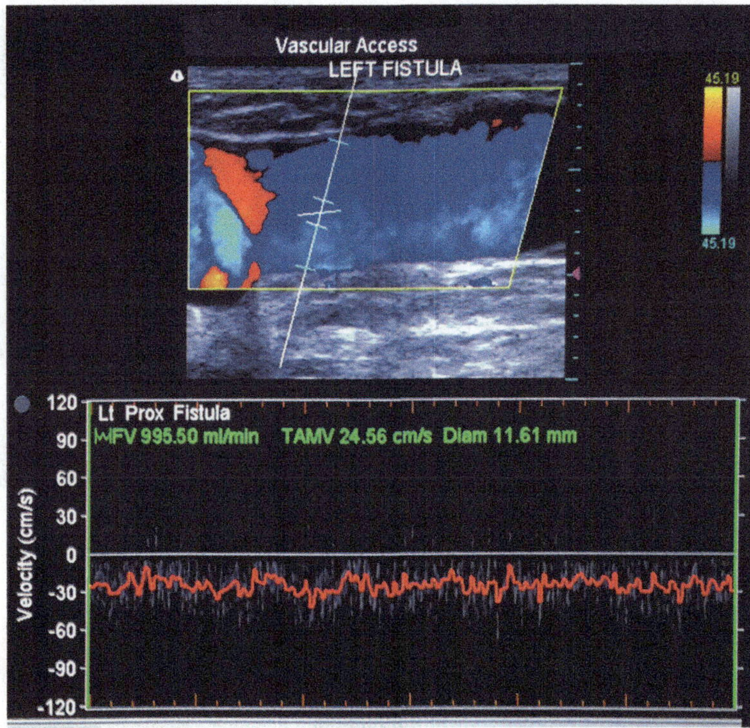

Fig. 11.5 Ultrasound determined access blood flow of 995.50 mL/min

Fig. 11.6 Post-procedure angiogram. *A* site of banding

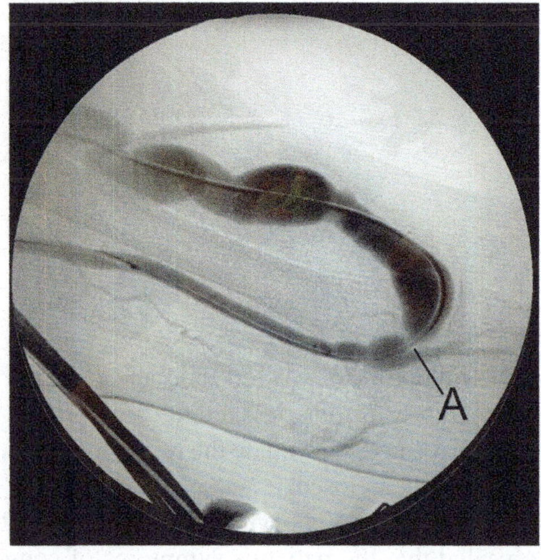

Discussion

It is well known that a dialysis vascular access with an inappropriately high-flow rate may be the cause of high-output heart failure [1–4]. This condition is defined as symptoms of cardiac failure – dyspnea either at rest or with varying degrees of exertion, orthopnea, paroxysmal dyspnea, and edema, either pulmonary and/or peripheral in the presence of high cardiac output. Specific characteristics of either the patient or the fistula, or both, may predispose to the development of heart failure. Some suggest that cardiac decompensation due to an arteriovenous fistula (AVF) is likely to only occur in a patient with underlying cardiac disease [5].

This is an example of a case in which high access blood flow resulted in aggravation of the patient's cardiac problems. The first question that should be asked is "how high is too high?" Currently there are no guidelines to answer this question. Two metrics have been used in efforts directed toward addressing this issue – access blood flow (Qa) and cardiopulmonary recirculation (CPR) which is defined as the ratio of Qa to cardiac output (CO). The Vascular Access Society guidelines [6] defined high dialysis access blood flow as one that is greater than 1.0 to 1.5 L/min or a cardiopulmonary resuscitation (CPR) greater than 20%. Most reports in the literature dealing with congestive heart failure secondary to high Qa have intervened in patients having a Qa equal to or greater than 2 L/min or a CPR of 30 to 35% [7]. Another metric that had been used is the cardiac index. The normal range for this metric in patients at rest is 2.6 to 4.2 L/min/m². A patient with a cardiac index greater than to 4.2 L/min/m² has a high cardiac output by definition.

The first approach in dealing with a dialysis patient who is having signs and symptoms of congestive heart failure is to eliminate contributing factors such as high blood pressure, anemia, and fluid overload. Once these problems have been eliminated, if the signs and symptoms of congestive heart failure process attention should be directed toward the arteriovenous access. In this setting, one access should be closed immediately. Short of this possibility, it is reasonable to intervene if the dialysis access has a Qa equal to or greater than 2.0 L/min or a CPR equal to or greater than 30% [8].

It has been suggested that hemodialysis patients be screened for potential high-output failure using the Qa/CO ratio; those patients with high access flows and Qa/CO ratios equal to or greater than 30% should undergo regular biannual echocardiographic assessment including left ventricular (LV) end-diastolic and systolic dimensions, LV mass index, and ejection fraction. If the patients with elevated Qa/CO ratios have increasing LV cavity volume and CO, then they may be considered for a fistula flow reduction [9].

Several surgical techniques have been proposed [10] for the treatment of high-output heart failure; all are based on an attempt to increase resistance at the level of the anastomosis or of the venous outflow in order to obtain a reduction of Qa. Balloon-assisted banding, a technique first described in 1999 [11], was used in this case resulting in a decrease in Qa and resolution of the patient's symptoms.

References

1. Wytrzes L, Markley HG, Fisher M, Alfred HJ. Brachial neuropathy after brachial artery-antecubital vein shunts for chronic hemodialysis. Neurology. 1987;37(8):1398–400.
2. Redfern AB, Zimmerman NB. Neurologic and ischemic complications of upper extremity vascular access for dialysis. J Hand Surg Am. 1995;20(2):199–204.
3. Isoda S, Kajiwara H, Kondo J, Matsumoto A. Banding a hemodialysis arteriovenous fistula to decrease blood flow and resolve high output cardiac failure: report of a case. Surg Today. 1994;24(8):734–6.
4. Young PR Jr, Rohr MS, Marterre WF Jr. High-output cardiac failure secondary to a brachiocephalic arteriovenous hemodialysis fistula: two cases. Am Surg. 1998;64(3):239–41.
5. London GM. Left ventricular alterations and end-stage renal disease. Nephrol Dial Transplant. 2002;17(Suppl 1):29–36.
6. Vascular Access Society. Guideline 20.3: Management of High Flow in AV Fistula and Graft. www.vascularaccesssociety.org.
7. Basile C, Lomonte C, Vernaglione L, Casucci F, Antonelli M, Losurdo N. The relationship between the flow of arteriovenous fistula and cardiac output in haemodialysis patients. Nephrol Dial Transplant. 2008;23(1):282–7.
8. Wasse H, Singapuri MS. High-output heart failure: how to define it, when to treat it, and how to treat it. Semin Nephrol. 2012;32(6):551–7.
9. MacRae JM, Pandeya S, Humen DP, Krivitski N, Lindsay RM. Arteriovenous fistula-associated high-output cardiac failure: a review of mechanisms. Am J Kidney Dis. 2004;43(5):e17–22.
10. Katz S, Kohl RD. The treatment of hand ischemia by arterial ligation and upper extremity bypass after angioaccess surgery. J Am Coll Surg. 1996;183(3):239–42.
11. Beathard GA, Settle SM, Shields MW. Salvage of the nonfunctioning arteriovenous fistula. Am J Kidney Dis. 1999;33(5):910–6.

Chapter 12
Ruptured Angioplasty Balloon

Gerald A. Beathard

Case Presentation

The patient was a 54-year-old male with an upper arm arteriovenous graft (AVG). The patient was referred to the dialysis access facility because of poor blood flow detected at the dialysis clinic. On physical examination, the patient had a hyperpulsatile AVG with a strong thrill and a loud bruit at the venous anastomosis, which was systolic only. After the patient was prepped and draped, an angiogram was performed which showed 80% stenosis of the venous anastomosis. This was treated using an 8 × 4 angioplasty balloon, which was advanced over a guidewire that had been inserted through a 7-Fr sheath. The lesion proved to be very resistant to dilatation. During the attempts at dilatation, the angioplasty balloon ruptured. Difficulty was encountered in extracting the angioplasty balloon through the sheath and required considerable force to complete the removal. Once extracted, examination of the ruptured angioplasty balloon revealed that the rupture had been circumferential and the front segment of the balloon was missing (Fig. 12.1). An angiogram was performed and showed a defect just proximal to the venous anastomosis (Fig. 12.2).

The primary guidewire was left in place. A second guidewire was inserted through the sheath and advanced beyond the site of the balloon fragment. The sheath was then removed and reinserted only over the second guidewire, excluding the primary guidewire. A 5-Fr vascular catheter was advanced through the sheath over the guidewire to a point several centimeters beyond the site of the balloon fragment. A snare (6–10 mm EnSnare®, Merit Medical Systems, Inc., South

G.A. Beathard, MD, PhD (✉)
Lifetime Vascular Access, University of Texas Medical Branch,
5135 Holly Terrace, Houston, TX 77056, USA
e-mail: gbeathard@msn.com

© Springer International Publishing AG 2017
A.S. Yevzlin et al. (eds.), *Dialysis Access Cases*,
DOI 10.1007/978-3-319-57500-1_12

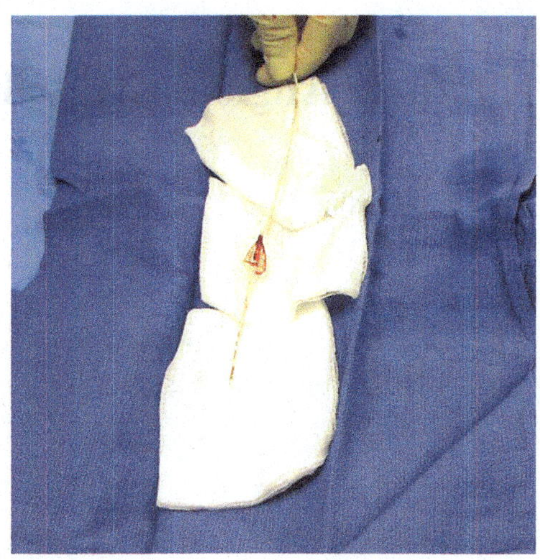

Fig. 12.1 Appearance of angioplasty balloon after extraction. Note that half of balloon is missing

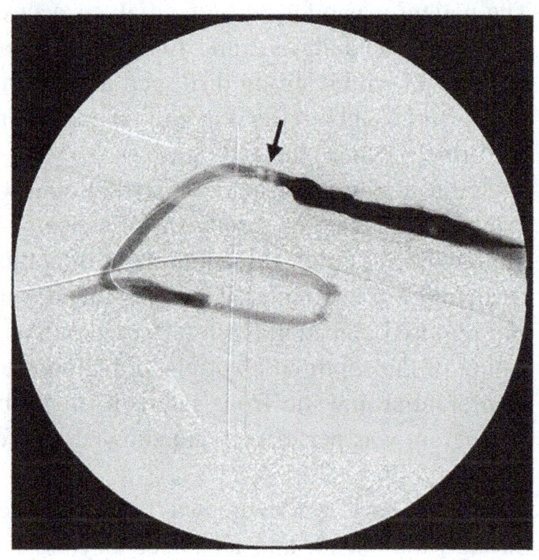

Fig. 12.2 Angiogram showing location of angioplasty balloon fragment on guidewire (*arrow*)

Jordan, UT, USA) was then advanced and deployed beyond the tip of the catheter. The tip of the primary guidewire was retracted to a point below the position of the snare being careful not to go beyond the site of the balloon fragment. Once below the position of the snare, the guidewire was advanced so that it would pass through one of the loops of the snare (Fig. 12.3). In this position, the guidewire was snared and brought out through the sheath. With both ends of the primary guidewire in hand, it was extracted along with the sheath, retrieving the balloon fragment (Fig. 12.4). Hemostasis was then obtained and the patient was discharged.

Fig. 12.3 Snaring primary
guidewire

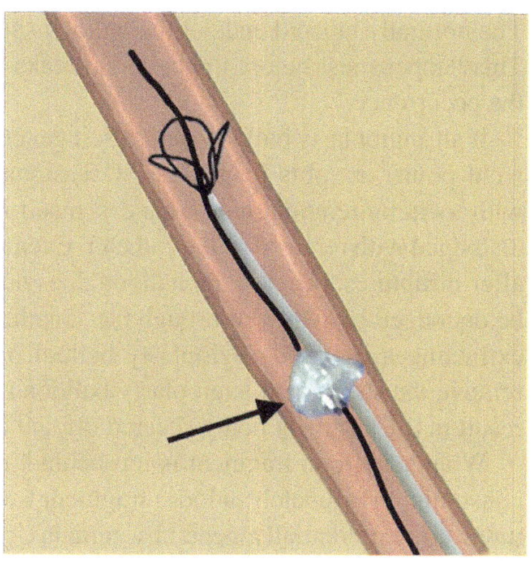

Fig. 12.4 Angioplasty
balloon fragment after
retrieval

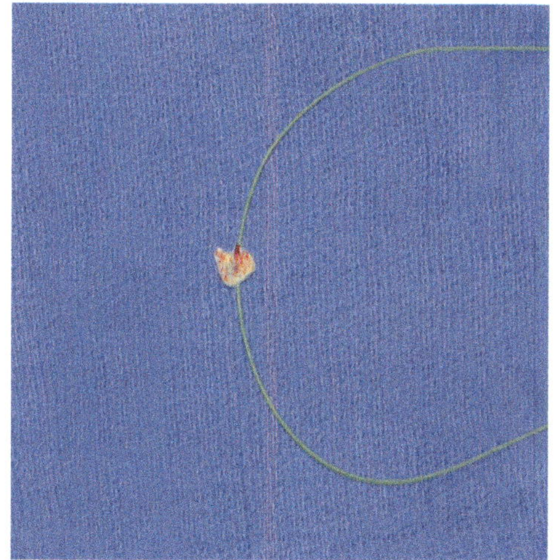

Discussion

Dilatation-resistant lesions are not uncommon, and some types of lesion are notorious
for being resistant. One of these is the venous anastomosis of an AVG. All angioplasty balloons have a pressure rating listed on the package, listed as rated burst
pressure in atmospheres. Although there is some safety margin in this stated value
with increasing pressure being applied, the balloon will eventually rupture.

Although rupturing an angioplasty balloon from excessive pressure should not
create a problem, this is not always the case. Therefore, rupture should be avoided
if possible. With some balloon designs, an impending rupture can be recognized.

The normally tapered ends of the balloon shoulder become rounded like a sausage. This happens just before the balloon breaks. Its recognition can be used to prevent the occurrence.

If an angioplasty balloon ruptures, it generally simply comes loose at the end (the weld point) or splits lengthwise. Occasionally, however, the balloon can rupture with a circumferential tear. This can result in a segment of the balloon becoming dislodged with removal of the catheter. Even if the fragment is still partially attached after it ruptures, because the balloon does not completely deflate, the fragment can be dislodged by removal through the sheath. For this reason, if resistance is met in extracting a ruptured angioplasty balloon from the sheath, one should stop and remove the sheath and angioplasty balloon together as a unit. This is less likely to result in the fragment being sheared off and left behind.

When a balloon fragment is left behind, retrieval is necessary and can be time-consuming. Fortunately, a loose fragment from an angioplasty balloon that has ruptured circumferentially generally remains on the guidewire. For this reason, the guidewire should always be left in place when an angioplasty balloon ruptures until the situation can be evaluated.

The fact that the fragment is still on the guidewire (assuming that the guidewire has been left in place) greatly facilitates its retrieval. By using an intravascular snare as described above, the leading end of the guidewire can be captured and extracted so that both ends are in hand with the fragment on the loop that has been created. By extracting the guidewire loop, the fragment can be retrieved easily.

If the guidewire is inadvertently removed prior to retrieval of the fragment, the task becomes much more difficult. Attempts at capturing it with a snare are complicated by the fact that it is not radiopaque. This makes localization much more difficult.

Chapter 13
Arteriovenous Fistula Percutaneous Flow Reduction: Balloon-Assisted Banding

Alexander S. Yevzlin

Case Presentation

A 72-year-old woman with a left upper-arm arteriovenous fistula (AVF) was referred for evaluation of left hand pain on hemodialysis. A detailed history and physical examination ruled out the possibility of nonischemic hand pain due to carpal tunnel syndrome or neuropathy. Preoperative evaluation revealed poor distal pulses in the left hand, as well as the contralateral arm.

An imaging catheter was inserted through the AVF and advanced into the left brachial artery. Angiography showed excellent flow of contrast into the AVF but minimal flow into the left ulnar and radial arteries (Fig. 13.1). This was suggestive of "true" steal syndrome, but further imaging of the arterial tree was obtained to rule out a focal arterial lesion, diffuse arterial disease, or thrombosed distal brachial artery. Specifically, two imaging studies were performed: (1) an assessment of inflow from the aortic arch (Fig. 13.2) and (2) a direct injection of contrast into the brachial artery while manually occluding the AVF (Fig. 13.3). Figure 13.4 shows restoration of flow into diminutive left radial and ulnar arteries with AVF occlusion.

Instead of ligation or surgical banding, a balloon-assisted percutaneous flow reduction was performed as previously described [1, 2]. Briefly, the brachial artery diameter was measured angiographically, and the same-size (4 mm) angioplasty balloon was then inflated to 10 atm pressure with complete effacement within the peri-anastomotic segment of the AVF. The balloon was percutaneously palpated, and after lidocaine injection, a 2-cm transverse incision was made over the middle of the balloon. Soft-tissue dissection was performed to expose the fistula with the

A.S. Yevzlin, MD, FASN (✉)
University of Michigan, Ann Arbor, MI, USA
e-mail: asy@medicine.wisc.edu

© Springer International Publishing AG 2017
A.S. Yevzlin et al. (eds.), *Dialysis Access Cases*,
DOI 10.1007/978-3-319-57500-1_13

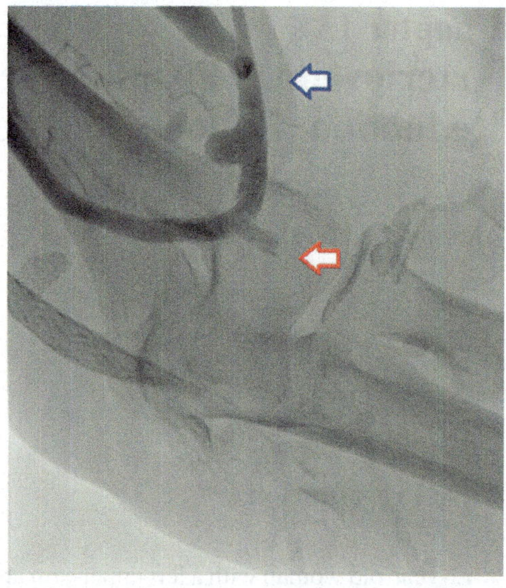

Fig. 13.1 Direct angiogram shows excellent flow of contrast into the arteriovenous fistula (AVF – *blue arrow*) but minimal flow into the distal brachial artery/ arteries (*red arrow*)

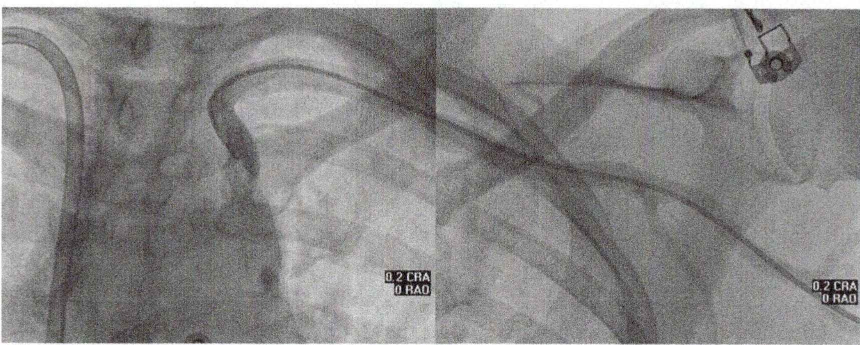

Fig. 13.2 Arterial angiogram from the subclavian artery

inflated balloon in it. Soft-tissue dissection was performed beneath the fistula, and a 2-0 silk thread was then tied around the balloon. The balloon was deflated and a post-procedure angiogram was obtained (Fig. 13.4).

Discussion

Hand ischemia in the setting of an arteriovenous access on the ipsilateral side was for many years referred to simply as steal syndrome. Although steal syndrome is an important clinical entity, it is one of many potential causes of hand ischemia. Although most arteriovenous accesses show evidence of arterial steal, ischemic

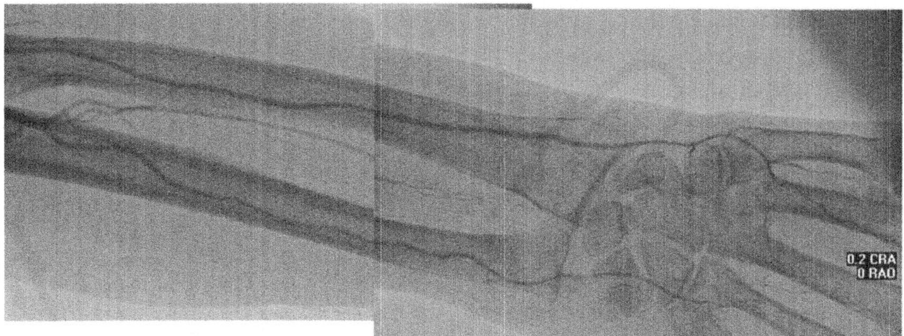

Fig. 13.3 Direct injection of contrast into the brachial artery (a) while manually occluding the arteriovenous fistula (AVF) shows restoration of flow into diminutive radial and ulnar arteries

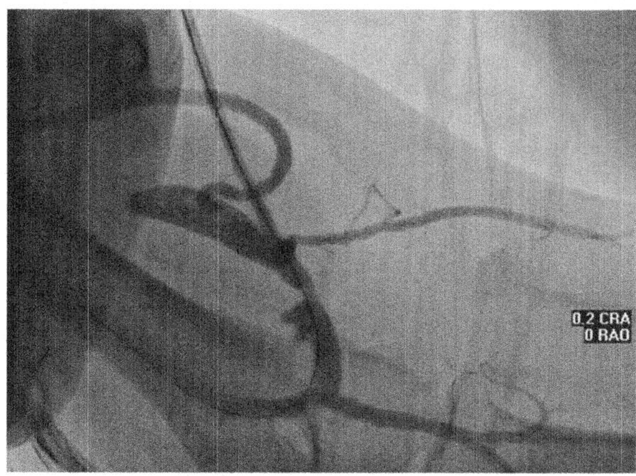

Fig. 13.4 Restoration of flow into the radial and ulnar vessels was achieved with balloon-assisted flow reduction

symptoms are rarely found [3]. Based on this observation, one can conclude that peripheral hypoperfusion and ischemia are more fundamental problems than arterial steal per se.

Distal hypoperfusion ischemia syndrome (DHIS) can occur in two specific entities that should not be described as steal syndrome: (1) focal arterial disease and (2) diffuse distal arteriopathy. Previous reports have highlighted that substantial ($\geq 50\%$) arterial stenoses are common in dialysis patients who present with symptoms of hand ischemia or vascular access dysfunction [4, 5]. Such lesions may be present anywhere within the arteries of the upper extremities, including the proximal arteries. Diffuse arteriopathy caused by vascular calcification and diabetes may also be a key contributor to the evolution of symptoms of distal hypoperfusion ischemia syndrome [6].

True steal syndrome can only be diagnosed if the aforementioned clinical scenarios are ruled out. The syndrome is defined by high blood flow volume through an arteriovenous anastomosis that steals blood from the forearm arteries, which otherwise have no pathology. In the setting of true steal syndrome, arteriovenous access ligation has historically been the treatment of choice [3]. Such an approach leads to loss of an arteriovenous access.

References

1. Salman L, Maya ID, Asif A. Current concepts in the pathophysiology and management of arteriovenous access-induced hand ischemia. Adv Chronic Kidney Dis. 2009;16:371–7.
2. Miller GA, Goel N, Friedman A, et al. The MILLER banding procedure is an effective method for treating dialysis-associated steal syndrome. Kidney Int. 2010;77:359–66.
3. Asif A, Leon C, Merrill D, et al. Arterial steal syndrome: a modest proposal for an old paradigm. Am J Kidney Dis. 2006;48:88–97.
4. Valji K, Hye RJ, Roberts AC, Oglevie SB, Ziegler T, Bookstein JJ. Hand ischemia in patients with hemodialysis access grafts: angiographic diagnosis and treatment. Radiology. 1995;196:697–701.
5. Leon C, Asif A. Arteriovenous access and hand pain: the distal hypoperfusion ischemic syndrome. Clin J Am Soc Nephrol. 2007;2:175–83.
6. Goldsmith DJ, Covic A, Sambrook PA, Ackrill P. Vascular calcification in long-term haemodialysis patients in a single unit: a retrospective analysis. Nephron. 1997;77:37–43.

Chapter 14
Hemodynamic Monitoring of Arteriovenous Fistulagram

Joel E. Rosenberg and Alexander S. Yevzlin

Case Presentation

A 75-year-old Caucasian woman with a history of chronic kidney disease stage IV after a deceased donor renal transplant due to polycystic kidney disease was admitted for a mitral valve replacement. Her severe mitral regurgitation was presumed to be secondary to rheumatic heart disease.

Her vascular access consisted of a right brachial artery to basilic vein fistula, and there is no evidence of any hemodynamically significant stenosis, although multiple areas of dilation with tortuosity limited the ability to accurately calculate a volume flow rate. She had mild aortic stenosis and regurgitation, severe mitral regurgitation, and mildly elevated pulmonary artery systolic pressures. Catheterization was notable for an ejection fraction of 35%, left ventricular end-diastolic artery pressure of 12 mmHg, systolic pulmonary artery pressure of 42 mmHg, pulmonary capillary wedge pressure (PCWP) of 19 mmHg, and a CO of 5 l/min with no significant obstructive coronary artery disease present.

Given the valvular heart disease, it was difficult to clinically evaluate the effect of the arteriovenous fistula (AVF) on the patient's heart failure. The CO remained below 8 l/mi, but a significant volume of forward output was lost secondary to the severe mitral regurgitation. A duplex of the fistula showed a seemingly severe increased flow with 12.1 l/min.

After the usual prep and drape, the AVF was accessed using a 7 Fr introducer. The fistulagram revealed no outflow abnormalities (Fig. 14.1). A pulmonary artery

J.E. Rosenberg, MD
Division of Nephrology, Department of Medicine, University of Wisconsin School of Medicine and Public Health, Madison, WI, USA

A.S. Yevzlin, MD, FASN (✉)
University of Michigan, Ann Arbor, MI, USA
e-mail:asy@medicine.wisc.edu

© Springer International Publishing AG 2017
A.S. Yevzlin et al. (eds.), *Dialysis Access Cases*,
DOI 10.1007/978-3-319-57500-1_14

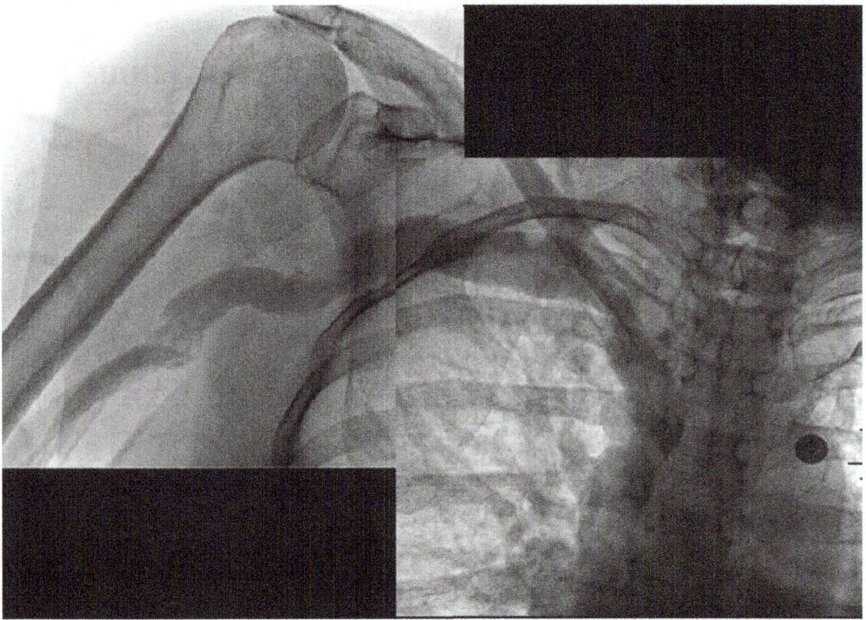

Fig. 14.1 Patent outflow of the AVF to the central vasculature

Fig. 14.2 Pulmonary
artery catheter inserted via
AVF. *Red arrow* points to
inflated (wedged) balloon

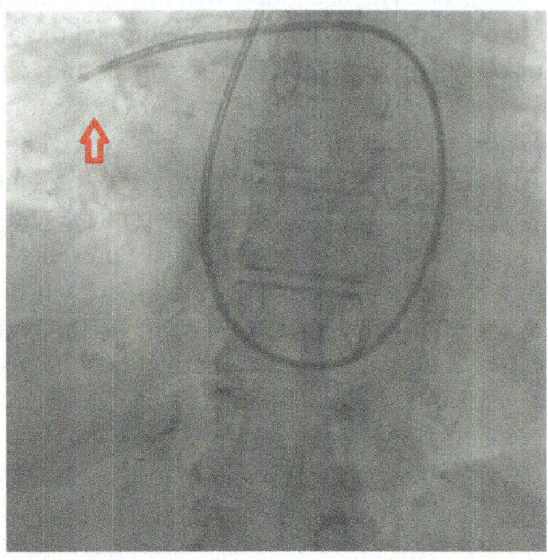

catheter was inserted into the AVF and passed directly into the pulmonary vasculature (Fig. 14.2). The patient's baseline cardiac output was 4.6 l/min, and she continued to have mildly elevated PCWP of 20 mmHg, PAP of 20 mmHg, and right atrial pressure of 10 mmHg. After occluding the AV fistula for 7 min, a 15% reduction of cardiac output to 3.9 l/min was noted, along with PCWP of 23 mmHg, PAP of 42/25 mmHg, and right atrial pressure of 18 mmHg.

Discussion

The minimal decrease in cardiac output with occlusion of the fistula and evidence suggested that the AVF was not contributing to the patient's heart failure. Intriguingly, she did have multiple areas of dilation with tortuosity generating a natural flow reduction. The decision was made to proceed with the mitral valve replacement but leave the AVF in place.

The exact incidence of high-output heart failure in ESRD patients is unknown, but heart disease, male sex, location of vascular access (proximal more than distal), and an upper arm fistula in the same arm as a previously functioning lower arm fistula are associated with increased risk [1]. Increased vascular access blood flow is a risk factor as well, but there is currently no consensus on the definition of high-flow AVG and AVFs because most of the data comes from case reports. The potential effect of an AVF on cardiac function is a matter of great concern to cardiologists and nephrologists alike. Rather than conjecture, real-time hemodynamic testing at the time of vascular access intervention can be performed [2].

References

1. Anand IS, Florea VG. High output cardiac failure. Curr Treat Options Cardiovasc Med. 2001;3:151–9.
2. Singh S, Elramah M, Allana SS, Babcock M, Keevil JG, Johnson MR, Yevzlin AS, Chan MR. A case series of real-time hemodynamic assessment of high output heart failure as a complication of arteriovenous access in dialysis patients. Semin Dial. 2014;27(6):633–8.

Chapter 15
Arterial Embolus

Gerald A. Beathard

Case Presentation

The patient was a 54-year-old male with end-stage renal disease secondary to polycystic kidney disease. The patient had a left brachial cephalic fistula that was 3 years old. He was referred to the vascular access center with a diagnosis of thrombosed fistula. Physical examination of his fistula revealed the absence of a pulse and thrill. Cannulation sites were evident indicating the course of the fistula; however, the fistula itself was not palpable.

Using ultrasound guidance, the lower fistula just above the anastomosis was cannulated in an antegrade direction; a guidewire was inserted followed by the insertion of a 7 F sheath. A 5 F vascular catheter was inserted over the guidewire and passed up to the level of the central veins. Radiocontrast was injected revealing that the central veins were patent with no evident pathology. Radiocontrast was slowly injected as the catheter was withdrawn from the fistula. This revealed that the lumen of the fistula was very narrow (Fig. 15.1). Only a small amount of thrombus was present.

After the outflow of the fistula was occluded using a sterile tourniquet, 2 mg of TPA (Cathflo®, alteplase) with saline to give a 5 ml final volume was injected through the 5 F vascular catheter throughout the length of the fistula. After a 5-min waiting period, an 8 × 4 angioplasty balloon was inserted in the upper portion of the AVF which was then dilated (Fig. 15.2). The AVF was then cannulated a second time in the upper portion with the needle pointed retrograde and a second sheath was inserted. This site was used to treat the lower part of the fistula with the angioplasty balloon (Fig. 15.3).

G.A. Beathard, MD, PhD (✉)
Lifetime Vascular Access, University of Texas Medical Branch,
5135 Holly Terrace, Houston, TX 77056, USA
e-mail: gbeathard@msn.com

© Springer International Publishing AG 2017
A.S. Yevzlin et al. (eds.), *Dialysis Access Cases*,
DOI 10.1007/978-3-319-57500-1_15

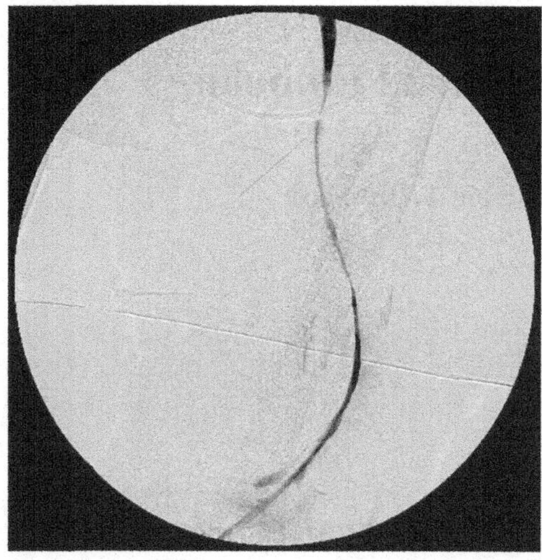

Fig. 15.1 Angiogram of thrombosed AVF. The lumen appears to be extremely narrowed; only a small amount of thrombus is present

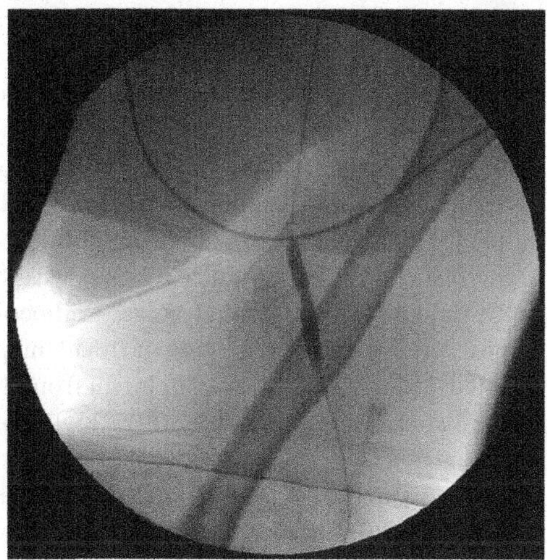

Fig. 15.2 Angioplasty of the upper portion of the AVF

Following angioplasty, flow was restored to the fistula. However, the patient began to complain of hand pain. Examination of the patient's hand revealed that it was cold and cyanotic. The opposite hand was warm with normal coloration. Additionally, the radial pulse, which had been present prior to the procedure, was no longer present.

The guidewire was inserted through the proximal sheath and passed downward through the fistula, across the anastomosis, and into the brachial artery. An arteriogram was performed at this point which revealed the presence of a filling defect in the proximal radial artery (Fig. 15.4). The vascular catheter was removed, and the

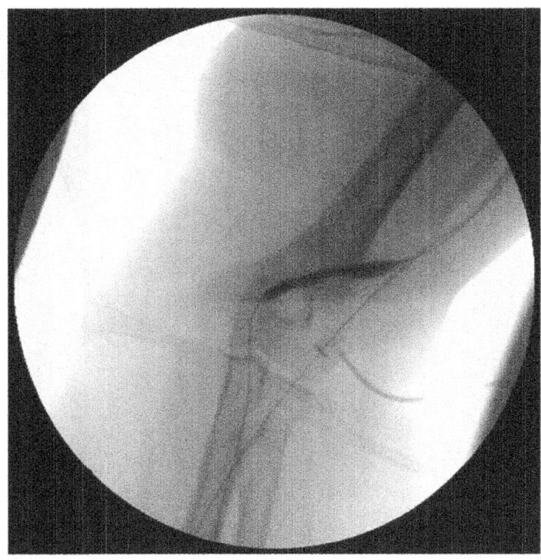

Fig. 15.3 Angioplasty of the lower portion of the AVF

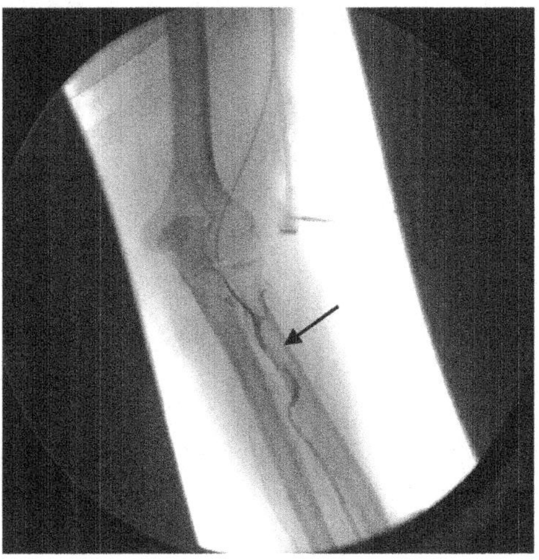

Fig. 15.4 Arteriogram of the lower brachial artery and radial artery. Embolus is present in proximal radial artery (*arrow*)

guidewire was advanced beyond the level of the embolus. A 4 × 4 angioplasty balloon was passed over the guidewire to a level beyond the embolus. The balloon was inflated and withdrawn back into the fistula. The vascular catheter was reinserted; radial contrast was injected to obtain an arteriogram (Figs. 15.5 and 15.6). This showed good flow in both the brachial and radial arteries.

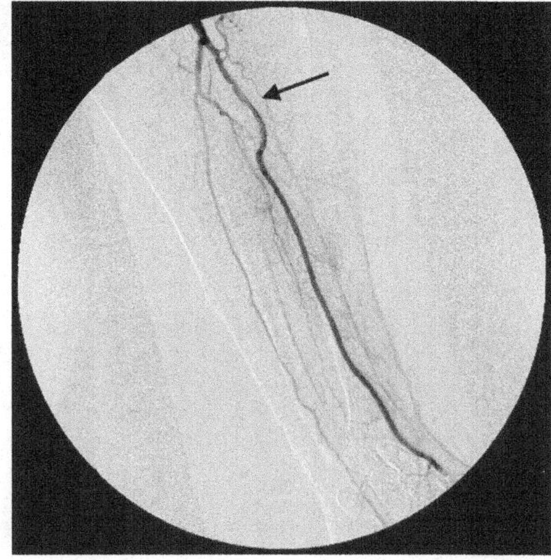

Fig. 15.5 Arteriogram post embolectomy. Good flow in the radial artery is now present, site of previous embolus indicated by *arrow*

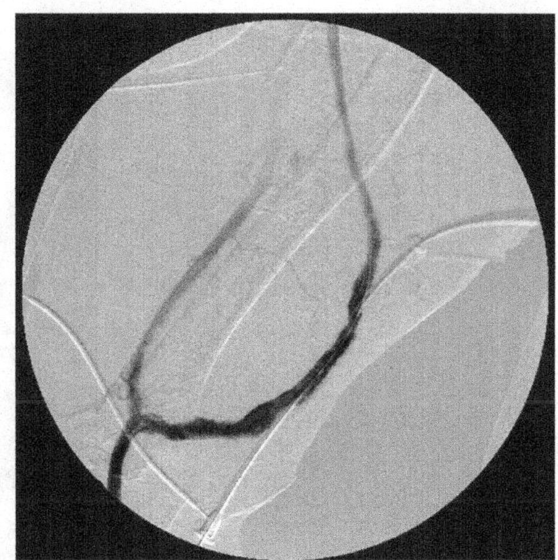

Fig. 15.6 Angiogram of AVF showing good blood flow

Discussion

In this AVF thrombectomy case, there are several points in the procedure that could have resulted in arterial embolization. Any injection of radial contrast into a fistula that is not flowing creates the risk of retrograde flow pushing thrombus across the anastomosis into the artery. The manipulation of a device such as an angioplasty balloon within the fistula can result in embolization. Additionally, there are instances in which embolization occurs without any obvious explanation.

Although they can occur with either, arterial emboli are more commonly associated with the thrombectomy of a graft than a fistula [1–7]. The reasons for this are not clear. It should be noted that the thrombus present in a fistula is mildly inflammatory and tends to become attached to the vessel wall. This decreases its ability to become detached and embolize.

The symptoms of embolization are those of hand ischemia. The hand and especially the fingers turn cold and take on a bluish discoloration that becomes mottled. These changes generally come on with the sudden onset of pain. In evaluating a patient's hand for a suspected embolus, it is important to compare it with the opposite hand. If both are cold and mottled, it is not likely that the hand in question reflects an acute problem. The pulses at the wrist are generally absent or considerably diminished, a change that can be appreciated only if the patient was carefully evaluated prior to having the thrombectomy procedure. A Doppler signal is generally present over the arteries at the wrist even when the pulse is not palpable, although it is frequently diminished. If nothing is detected with Doppler examination, the urgency for immediate treatment to avoid tissue damage is even greater than usual.

As is the case with all procedure-related complications, the first aspect of management is avoidance. Although the occurrence of small asymptomatic and therefore inconsequential emboli may be unavoidable when doing a thrombectomy procedure, it is important to take measures to avoid the introduction of large clot fragments across the arterial anastomosis. Fluids (saline, radiocontrast, medications) are commonly introduced into the access during a thrombectomy procedure; care must be exerted to avoid doing it too rapidly and never doing it if the outflow is obstructed. If the outflow is not open, it will generally go retrograde. Care must be taken even after the thrombectomy procedure is completed and the angiogram appears pristinely clean. One should never occlude the access and do a retrograde injection to visualize the anastomosis and adjacent artery following the thrombectomy. Even if the access looks clean angiographically, small clot fragments will still be present. Additionally, care must be used in passing devices across the arterial anastomosis during the thrombectomy procedure. It is possible to push material into the artery resulting in a problem.

Symptomatic emboli must be treated in a timely fashion in order to prevent permanent sequelae. Treatment should be directed at restoring flow to the ischemic hand as quickly as possible in order to relieve the patient's pain and preserve hand function by avoiding secondary muscle ischemia and necrosis. Outcomes and prognosis largely depend on a rapid diagnosis and initiation of appropriate and effective therapy [6]. There are several approaches to the therapy of symptomatic peripheral artery emboli. These can be divided into endovascular and surgical (see Table 15.1). Further, the endovascular approach can be subdivided into mechanical and pharmacological.

Table 15.1 Treatment modalities for arterial emboli

Percutaneous – mechanical
Balloon catheter embolectomy [1, 2]
Catheter thromboaspiration [3, 4]
Back-bleeding [5]
Percutaneous – pharmacological
Thrombolysis [6]
Surgical embolectomy [6]

The endovascular techniques are generally effective. However, if they are not, patients with evidence of severe ischemia should not be treated with catheter-based thrombolytic therapy. This generally takes several hours, and threatened ischemic changes may become irreversible over the course of the treatment. These patients should be treated surgically on an emergent basis in order to avoid a permanent injury to the hand [6, 7]. Additionally, the thrombus that makes up the embolus may be chronic and, if so, is frequently somewhat resistant to fibrinolysis [8, 9].

References

1. Beathard GA. Management of complications of endovascular dialysis access procedures. Semin Dial. 2003;16:309.
2. Sofocleous CT, Schur I, Koh E, et al. Percutaneous treatment of complications occurring during hemodialysis graft recanalization. Eur J Radiol. 2003;47:237.
3. Sniderman KW, Bodner L, Saddekni S, et al. Percutaneous embolectomy by transcatheter aspiration. Work in progress. Radiology. 1984;150:357.
4. Turmel-Rodrigues LA, Beyssen B, Raynaud A, Sapoval M. Thromboaspiration to treat inadvertent arterial emboli during dialysis graft declotting. J Vasc Interv Radiol. 1998;9:849.
5. Trerotola SO, Johnson MS, Shah H, Namyslowski J. Backbleeding technique for treatment of arterial emboli resulting from dialysis graft thrombolysis. J Vasc Interv Radiol. 1998;9:141.
6. Rajan DK, Patel NH, Valji K, et al. Quality improvement guidelines for percutaneous management of acute limb ischemia. J Vasc Interv Radiol. 2005;16:585.
7. Beathard GA, Litchfield T, Physician Operators Forum of RMS Lifeline, Inc. Effectiveness and safety of dialysis vascular access procedures performed by interventional nephrologists. Kidney Int. 2004;66:1622.
8. Valji K, Bookstein JJ, Roberts AC, et al. Pulse-spray pharmacomechanical thrombolysis of thrombosed hemodialysis access grafts: long-term experience and comparison of original and current techniques. AJR Am J Roentgenol. 1995;164:1495.
9. Kumpe DA, Cohen MA. Angioplasty/thrombolytic treatment of failing and failed hemodialysis access sites: comparison with surgical treatment. Prog Cardiovasc Dis. 1992;34:263.

Part II
Arteriovenous Graft

Chapter 16
Exotic Arteriovenous Graft Creation

Paul A. Stahler and Robert R. Redfield III

Abbreviations

AVF	Arteriovenous fistula
SVC	Superior vena cava
HeRO	Hemodialysis Reliable Outflow
ePTFE	Expanded polytetrafluoroethylene
LEAVG	Lower extremity arteriovenous graft
TEVG	Tissue-engineered vascular graft
CWAVG	Chest wall arteriovenous graft

Case Presentation

The patient is a 38-year-old female with a history of end-stage renal failure secondary to type 1 diabetes mellitus on hemodialysis for 8 years. During this period of time, she has had difficulty maturing and maintaining multiple AVFs and required intermittent central venous catheters. Most recently, she was receiving hemodialysis via a right arm AVF with poor clearance. Previously performed angiograms demonstrated significant central stenosis with 100% occlusion at the right subclavian-SVC junction and 90% occlusion of the left innominate vein (Fig. 16.1a, b). The patient underwent successful balloon angioplasty with 30% residual stenosis of the left

P.A. Stahler, MD (✉)
Division of Transplant Surgery, Department of Surgery, Hennepin County Medical Center, 700 Park Ave, Minneapolis, 55415 Minnesota, United States
e-mail: stahler@surgery.wisc.edu

R.R. Redfield III, MD
Division of Transplant Surgery, Department of Surgery, University of Wisconsin School of Medicine and Public Health, Madison, WI, USA

© Springer International Publishing AG 2017
A.S. Yevzlin et al. (eds.), *Dialysis Access Cases*,
DOI 10.1007/978-3-319-57500-1_16

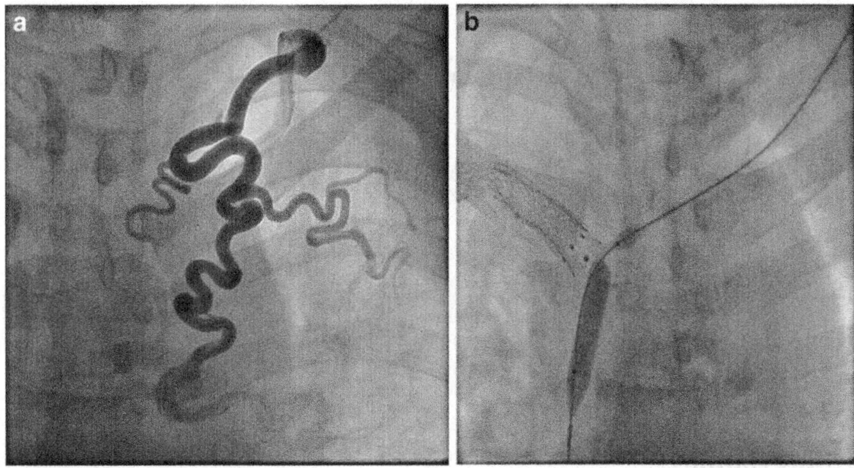

Fig. 16.1 (**a**) Angiogram of chest wall collaterals leading to a critical stenosis of the innominate vein. (**b**) Angioplasty of a 90% occlusion of the left innominate vein

innominate vein. She was subsequently referred to our center for additional dialysis access.

After reviewing her previous operative reports and imaging studies, she was taken to the operating room where a HeRO graft was placed. The left subclavian vein was accessed percutaneously, and the venous limb was positioned in the right atrium under fluoroscopic guidance and tunneled out to the deltopectoral groove. The arterial limb was anastomosed to the left brachial artery using an early cannulation Flixene graft which was then anastomosed to the ePTFE component of the HeRO graft. Subsequently the patient did well and had improved clearance on dialysis without the need for further central venous catheters.

Discussion

Hemodialysis Reliable Outflow Grafts

The HeRO graft (Hemosphere Inc., Minneapolis, MN) was developed for use in maintaining upper extremity vascular access for patients with central stenosis. The device is composed of a graft and a silicone catheter. The graft component is a 7-mm ePTFE graft with a titanium connector at one end. The venous outflow component is a 19F (outer diameter 6.3 mm, inner diameter 5 mm) nitinol-reinforced braid which prevents the catheter from kinking (Fig. 16.2). The end of the graft is typically anastomosed to the axillary or brachial artery and subsequently tunneled in the upper arm to a counter-incision near the deltopectoral groove. The venous outflow component is placed percutaneously into the internal jugular, subclavian, or large collateral veins and positioned at the level of the right atrium with the use of fluoroscopy. Similar to other grafts, the HeRO device is cannulated by inserting

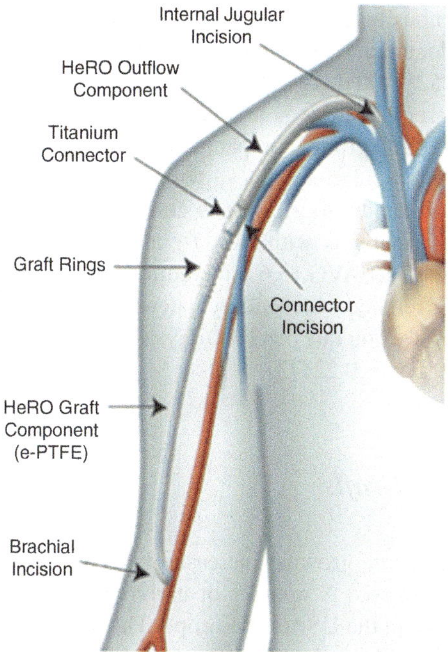

Fig. 16.2 Diagram of the Hemodialysis Reliable Outflow (HeRO) device is shown with the arterial anastomosis at the brachial artery near the antecubital fossa and the tip of the venous catheter position at the right atrium. The titanium connector links the graft and venous components. The device is inserted using a brachial incision in the proximal upper arm for the arterial anastomosis, an incision at the base of the neck for the initial venous cannulation, and an incision near the deltopectoral groove to facilitate passage of the venous component (Reproduced with permission from Glickman [1] ©Elsevier)

dialysis needles into the graft component. Immediate stick grafts can be successfully fashioned to the graft if needed.

Suitable patients for the HeRO device have typically had multiple tunneled dialysis catheters and failed upper extremity access. Preoperative evaluation of these patients includes a complete upper extremity duplex arterial and venous examination. A central venogram should also be performed to further define the venous anatomy prior to placement. Relative contraindications include brachial artery diameter <3 mm, congestive heart failure with an ejection fraction of <20%, active infections, and/or systolic blood pressure <100 mmHg [1].

Long-term outcomes of the HeRO device are lacking and are largely limited to single-center studies with short-term follow-up. Successful placement of the device has been reported in 95% of patients in several studies [2]. The primary and secondary patency rates in these studies range between 39–46% and 72–88%, respectively. The intervention rate was noted to be similar at 1.7–2.5 events/year compared to 1.6–2.4 events/year with prosthetic access.

Infection is a major complication of long-term catheter use for hemodialysis. Patients dependent on a catheter have a 1.6 times increased risk of sepsis compared to patients dialyzed through a fistula or graft. Although an existing catheter can be re-wired with the venous outflow component, an increased risk of infectious complications has been documented when the HeRO device was placed ipsilateral to the previous catheter (1.52/1000 days vs 0.3/1000 days) [2].

The overall bacteremia rate was significantly lower in the Katzman study at 0.7/1000 days in the HeRO group, compared to 1.7/1000 days in the catheter group. Frequently this patient population is faced with two options, HeRO placement and creation of a lower extremity arteriovenous graft (LEAVG). In a study comparing HeRO versus LEAVG, patients with HeRO grafts required more interventions to maintain patency versus LEAVG (2.6 interventions/year vs 1.7 interventions/year, $p = 0.003$) with no difference in secondary patency (77% vs 83%, $p = 0.14$). There was no difference in infection rate per 1000 days (0.61 vs 0.71, $p = 0.77$) or 6-month mortality (22% vs 19%, $p = 0.22$) [3].

Early Cannulation Grafts

Early cannulation grafts are growing increasingly as a popular option to avoid the use of catheters for dialysis. A number of early cannulation grafts are currently commercially available in the USA and Europe. The most extensively studied early cannulation graft in the USA is the Flixene graft (Maquet-Atrium Medical, Hudson, NH). The Flixene vascular graft is an ePTFE graft with a trilaminate composite structure. A hydrostatic protection membrane designed to minimize weeping and suture hole bleeding is interposed between a smooth inner blood-contacting surface designed to minimize platelet aggregation and a microporous outer anchoring surface designed to promote tissue adherence. In a prospective single-center nonrandomized study in which 82% of grafts were accessed within the first week after implantation, 6-month primary assisted and secondary patency rates were 65% and 86%, respectively, and 56% and 86% at 1 year [4]. Key to the success of these early cannulation grafts is education of dialysis nurses and staff regarding proper care and access to these grafts. The authors suggest marking the graft location, flow direction, and access sites (Fig. 16.3). In this single study, 1-year patency rates appeared to be comparable to those achieved with conventional grafts.

Novel Graft Technology

ePTFE is the most widely used graft for hemodialysis. A majority of these grafts eventually fail due to intimal hyperplasia at the venous anastomosis secondary to smooth muscle proliferation. Several attempts have been made to prevent graft thrombosis including systemic anticoagulation, antiplatelet therapy, and statins but have not shown real benefit in improving long-term patency. The Gore Propaten vascular graft (W.L. Gore & Associates, Inc., Flagstaff, AZ) was introduced for hemodialysis access in 2006. This graft contains heparin molecules covalently

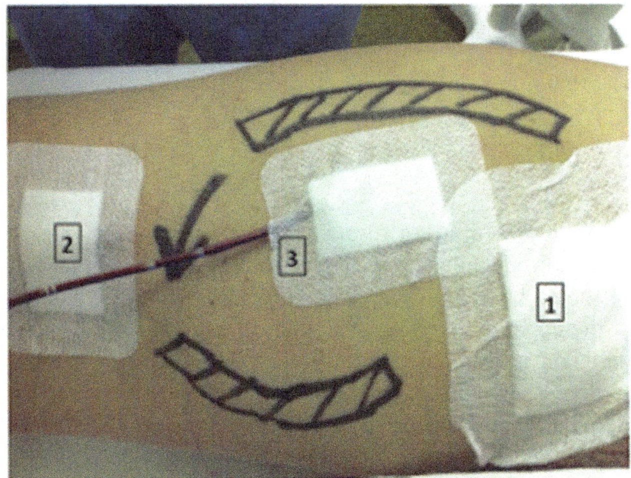

Fig. 16.3 Graft freshly implanted in the thigh with drawings showing potential cannulation sites (hatched) and flow direction (*arrow*). Dressings cover the cutdown for the femoral vessel approach (*1*), cutdown for loop tunnelization (*2*), and suction drain (*3*) (Reproduced with permission from Berard et al. [4] ©Elsevier)

bonded to the luminal surface providing a stable non-leaching heparin surface on the graft. Initial retrospective studies involving the Propaten graft showed mixed results in preventing graft thrombosis. A prospective, randomized study of Propaten versus standard ePTFE grafts showed no difference in primary patency at 6 and 12 months between Propaten and ePTFE (35% and 14% vs 29% and 12%, $p = 0.48$). Secondary patency was also similar between Propaten and ePTFE at 12, 24, and 36 months (83%, 83%, and 81% vs 81%, 73%, and 68%, $p = 0.33$). There were significantly fewer thromboses in the Propaten grafts during the first 5 months compared to standard grafts ($p = 0.02$) [5]. These results suggest a trend toward improved early patency but overall no significant difference in long-term patency.

The development of readily available tissue-engineered vascular grafts (TEVGs) is a promising area of research when a patient's own vasculature is not suitable for creation hemodialysis access. Initial experience involving TEVGs involved seeding autologous bone marrow cells or culturing autologous fibroblasts and endothelial cells with or without a scaffold. Major limitations of this technology include high production cost (>$15,000 per graft) and long wait times of up to 9 months. An alternative TEVG using allogeneic smooth muscle cells grown on a tubular polyglycolic acid has shown early promise (Humacyte, Inc., Morrisville, NC). In an early animal model, the Humacyte graft demonstrated similar mechanical properties to native human blood vessels with resistance to dilatation, calcification, intimal hyperplasia, and tolerated long-term storage at 4 °C. Long-term patency of these grafts was demonstrated up to 6 months with less intimal hyperplasia compared to standard PTFE grafts [6]. Additional advantages of this graft versus decellularized human cadaveric vessels include lack of side branches requiring ligation, creation of several grafts from a single donor, minimal immunogenicity, and short production time. Clinical trials of the Humacyte graft are ongoing.

Chest Wall Grafts

Chest wall grafts represent a last resort in patients with difficult upper extremity access. These are infrequently performed at rates less than 1% of all vascular access procedures. Patients being considered for CWAVG should undergo preoperative ultrasound of the upper extremities and proximal portions of the subclavian veins. A central venogram is recommended if high resistance indices are seen on ultrasound and/or there is a history of central stenosis.

The graft is placed by making 5–6-cm horizontal infraclavicular incision to access the axillary artery and vein. A PTFE graft is then tunneled in the subcutaneous tissue with a counter-incision at the inferior apex of the tunnel (Fig. 16.4). Proper tunneling of the graft is a technical challenge, and its location in relation to the nipple is key. The graft should course cephalad or caudal to the nipple rather than deep into it.

Patency outcomes in this challenging patient population are similar to other extremity grafts. One of the largest reported series of 67 grafts demonstrated that primary 1- and 2-year patency rates were 69.5% and 36.9% and secondary patency

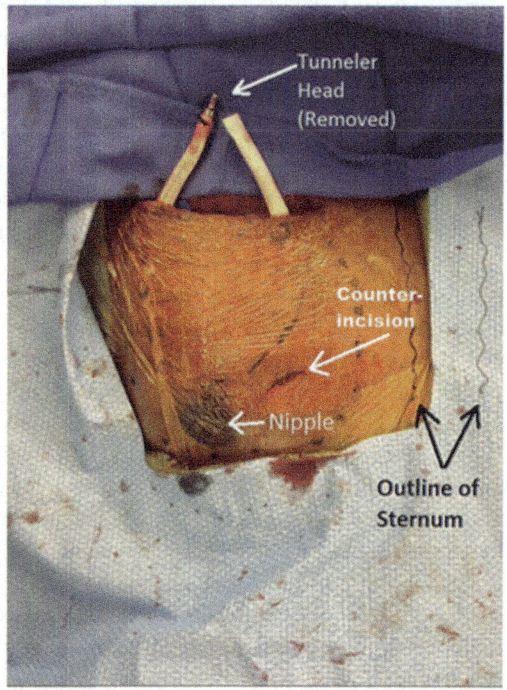

Fig. 16.4 Tunneled chest wall arteriovenous graft (CWAVG) with incision and counter-incision (Reproduced with permission from Liechty et al. [7] ©Elsevier)

rates were 81.6% and 57.6% [7]. Graft infections occurred in 9% of patients which is lower than rates reported for tunneled lines and femoral grafts, and lastly steal syndrome was not seen in any patient with a CWAVG. These results demonstrate the utility and durability of CWAVGs in this challenging patient population.

Pearls

- Preoperative evaluation of patients with a history of prior central venous catheters and failed arteriovenous access for dialysis should include a central venogram to evaluate for central stenosis. HeRO graft placement is a great option in patients with critical central stenosis.
- Early cannulation grafts can be safely accessed within 48 h and can potentially avoid the use of central venous catheters. Education of dialysis personnel is essential.
- Chest wall grafts remain a viable option at establishing long-term internal access in patients with a history of failed upper arm access and avoid the use of lower extremity arteriovenous grafts.

References

1. Glickman MH. HeRO vascular access device. Semin Vasc Surgery. 2011;24:108–12.
2. Katzman HE, McLafferty RB, Ross JR, et al. Initial experience and outcome of a new hemodialysis access device for catheter-dependent patients. J Vasc Surg. 2009;50:600–7.
3. Steerman SN, Wagner J, Higgins JA, et al. Outcomes comparison of HeRO and lower extremity arteriovenous grafts in patients with long-standing renal failure. J Vasc Surg. 2013;57: 776–83.
4. Berard X, Ottaviani N, Brizzi V, et al. Use of the Flixene vascular access graft as an early cannulation solution. J Vasc Surg. 2016;62:128–34.
5. Semesh D, Goldin I, Hijazi J, et al. A prospective randomized study of heparin-bonded graft (Propaten) versus standard graft in prosthetic arteriovenous access. J Vasc Surg. 2015;62: 115–2.
6. Dahl SL, Kypson AP, Lawson JH, et al. Readily available tissue-engineered vascular grafts. Sci Trans Med. 2011;3:68ra9.
7. Liechty JM, Fischer T, Davis W, et al. Experience with chest wall arteriovenous grafts in hemodialysis patients. Ann Vasc Surg. 2015;29:690–7.

Chapter 17
Accessory Brachial Artery Feeding Arteriovenous Graft

Gerald A. Beathard

Case Presentation

The patient was a 56-year-old female who had been on hemodialysis for slightly more than a year. The patient had a forearm loop arteriovenous graft (AVG) in her left arm. Her dialysis treatments were complicated by persistently marginal access blood flow and recurrent thrombosis, all of which had been treated surgically and had no imaging studies performed. Although the patient had never had vascular mapping performed, she had been told by the surgeon who placed the graft that she had poor veins and would probably need to have a central venous catheter for her dialysis treatments.

The patient was referred to the vascular access center because of another episode of AVG thrombosis. The standard AVG thrombectomy procedure was performed but was unsuccessful because of failure to restore inflow. Blood flow in the feeding artery was then evaluated, and it was found to be thrombosed. A vascular catheter guidewire combination was used to advance up the feeding artery and find the proximal extent of the thrombus. After going almost to the axilla, it was discovered that the feeding artery was not the brachial artery as expected but a smaller branch (Fig. 17.1). The AVG arterial anastomosis was with this vessel (Fig. 17.2). It was thought that this represented a high bifurcation of the brachial artery. However, further angiographic evaluation revealed that the bifurcation was slightly below the level of the elbow (Fig. 17.3). This confirmed that this vessel was actually an accessory brachial artery, which was feeding the AVG.

G.A. Beathard, MD, PhD (✉)
Lifetime Vascular Access, University of Texas Medical Branch,
5135 Holly Terrace, Houston, TX 77056, USA
e-mail: gbeathard@msn.com

© Springer International Publishing AG 2017
A.S. Yevzlin et al. (eds.), *Dialysis Access Cases*,
DOI 10.1007/978-3-319-57500-1_17

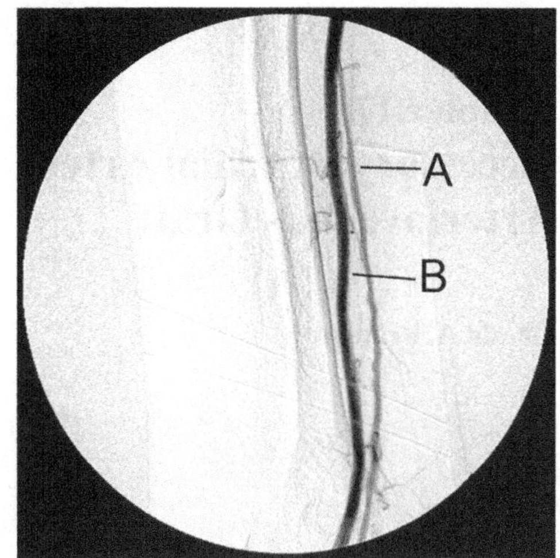

Fig. 17.1 Arteriogram showing accessory brachial artery (*A*) and brachial artery (*B*)

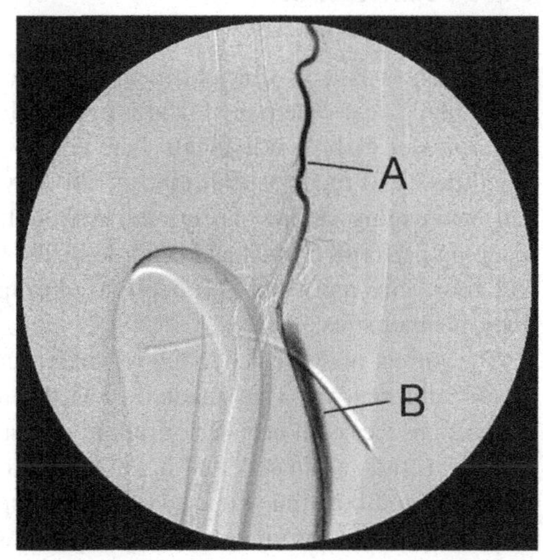

Fig. 17.2 Arteriogram showing the arterial anastomosis of the AVG. *A* accessory brachial artery, *B* AVG

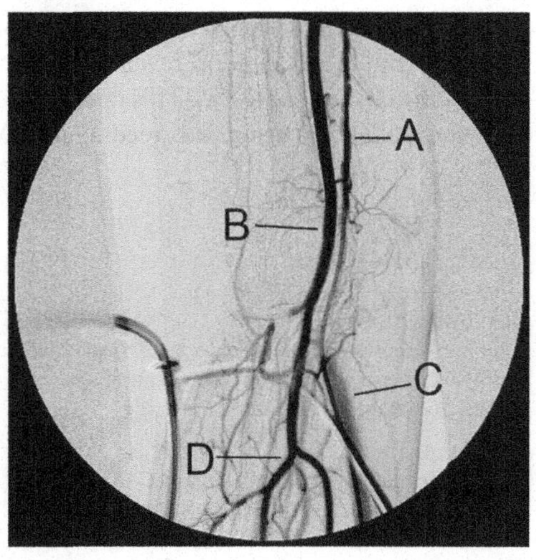

Fig. 17.3 Arteriogram showing the brachial artery bifurcation. *A* accessory brachial artery, *B* brachial artery, *C* AVG, *D* bifurcation of brachial artery

Discussion

The presence of an accessory brachial artery represents a variant in arterial anatomy associated with the brachial artery. The incidence of an this anatomical variant has been reported to be about 0.52% of patients or 1 in every 200 cases [1]. This accessory brachial artery originates from the upper third of the brachial artery, and passes downward to rejoin it proximal to the elbow, before the bifurcation. Both the brachial artery and the accessory brachial artery are closely related to the median nerve. However, nerve; however, their relationship is quite different. In the upper arm, the median nerve is immediately lateral to the brachial artery. Distally, the median nerve crosses the medial side of the brachial artery and lies somewhat anterior to it at the elbow joint. The accessory brachial artery lies lateral to the median nerve and crosses anterior to the nerve before it rejoins the brachial artery.

When what is assumed to be the brachial artery is examined at the time of surgical creation of an arteriovenous vascular access, the difference in its relationship to the median nerve in addition to its size being smaller than one would expect for the brachial artery should alert one to the fact that an accessory brachial artery is present.

When presented with a case such as this, the first consideration is generally that it is a high bifurcation, which is the more common variant. In a retrospective study of 220 dialysis patients who had preoperative duplex scans of their upper extremities, 26 (12%) patients were found to have high bifurcations of the brachial artery. Fourteen (53%) of these patients had bilateral scans. Of the 14 patients, 36% were noted to have high bifurcations bilaterally and 64% were unilateral.

If the accessory brachial artery is used in creation of the anastomosis for an AVG, two problems can occur. First, the small size of the artery very often provides less than optimum blood flow for the dialysis access. Second, since this artery is small and has no branches, the anastomosis that is created is basically an end-to-end connection. When blood flow ceases in the AVG thrombosis, flow also stops in the artery; this causes the thrombus to extend up the artery as it did in this case.

The problem being experienced by this patient, recurrent thrombosis, should be addressed by performing a surgical revision on the arterial anastomosis to connect the AVG to the true brachial artery.

Reference

1. Rodriguez-Niedenfuhr M, Vazquez T, Nearn L, Ferreira B, Parkin I, Sanudo JR. Variations of the arterial pattern in the upper limb revisited: a morphological and statistical study, with a review of the literature. J Anat. 2001;199(Pt 5):547–66.

Chapter 18
Swollen Arm with Suspected Arteriovenous Graft Infection

Gerald A. Beathard

Case Presentation

The patient was a 52-year-old male who started on dialysis 4 months ago with a tunneled-cuffed catheter. The catheter has been exchanged twice because of poor flow. A loop graft was placed in the patient's right arm forearm 10 days ago earlier. The patient was referred to the vascular access center because of swelling of his right arm and erythema over the course of the arteriovenous graft (AVG).

Examination of the patient revealed diffuse erythema over the course of the loop graft (Fig. 18.1). There was no fluctuance, no pain, and no exudation. The patient's right arm was moderately swollen and this was diffuse.

After the patient was prepped and draped, an angiogram was performed which showed the presence of stenosis within the superior vena cava (Fig. 18.2). A vascular catheter was introduced into the left brachiocephalic vein, and a second angiogram was performed to further define the anatomy of the central veins. This showed no abnormality other than the previously noted stenotic site within the superior vena cava (Fig. 18.3). A guidewire was advanced through the site and into the inferior vena cava. The stenotic site was dilated with a 12×4 angioplasty balloon with good result. A post-procedure angiogram performed at that time showed visualization of the spirit superior vena cava and also the azygos vein with radiocontrast outlining the cardiac and pulmonary venous structures (Fig. 18.4).

G.A. Beathard, MD, PhD (✉)
Lifetime Vascular Access, University of Texas Medical Branch,
5135 Holly Terrace, Houston, TX 77056, USA
e-mail: gbeathard@msn.com

© Springer International Publishing AG 2017
A.S. Yevzlin et al. (eds.), *Dialysis Access Cases*,
DOI 10.1007/978-3-319-57500-1_18

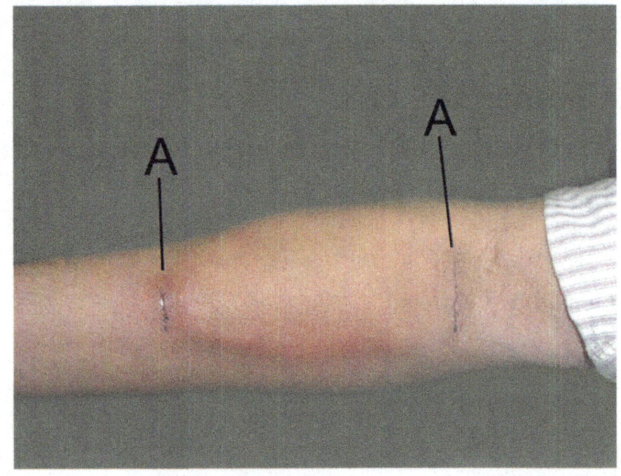

Fig. 18.1 Appearance of patient's right arm showing diffuse erythema over the entire course of the AVG. *A* Recently created incisions attesting to the fact that this access was recently placed

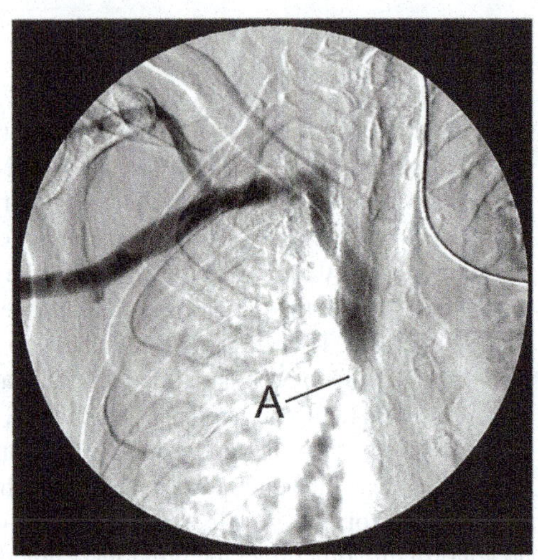

Fig. 18.2 Angiogram showing stenosis of superior vena cava. *A* Site of stenosis

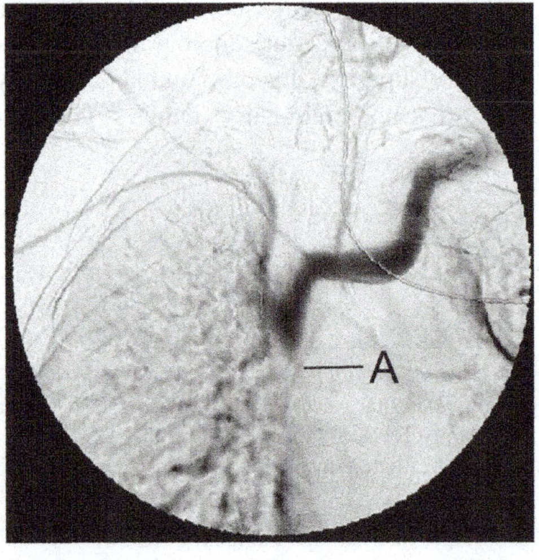

Fig. 18.3 Angiogram performed using vascular catheter showing left brachiocephalic vein and stenotic lesion in superior vena cava. *A* Site of stenosis

Fig. 18.4 Post-angioplasty angiogram showing good result from treatment. *A* Site of stenosis in superior vena cava, *B* azygos vein, *C* the right atrium, *D* pulmonary veins

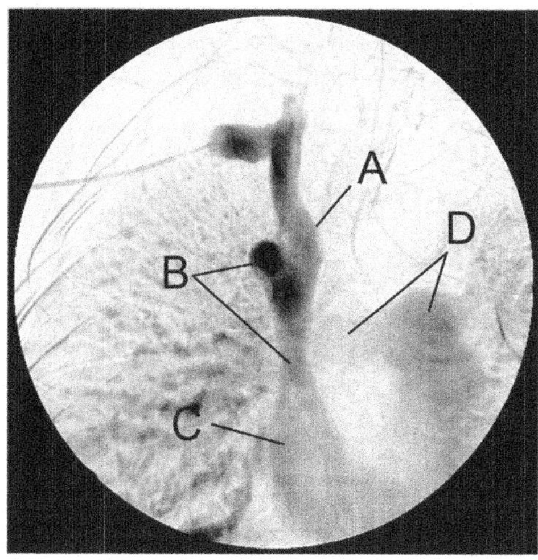

Discussion of Case

This case demonstrates two important points. First, diffuse erythema over the entire course of a recently placed synthetic graft should not automatically lead to the assumption that the graft is infected. This is typical of what has been referred to as a dermal flare. It is thought to be related to irritation of the dermis by the tunneling that took place when the graph was inserted. A deep infection of the graft can certainly occur at this time but is more likely to be localized and associated with other findings such as exudation and fluctuance.

Second, while some localized swelling may be observed postoperatively following the placement of a synthetic graft, diffuse swelling of the ipsilateral extremity is almost pathognomonic of a central vein stenosis. It is very unusual to observe this physical finding in the absence of such a lesion.

The prominence of the azygos vein on the post-angioplasty angiogram indicates that the obstruction caused by this lesion is at least partially compensated. The azygos system is the primary contributor to collateral flow from an obstruction in this area.

Chapter 19
Arteriovenous Graft Thrombosed for 3 Years

Gerald A. Beathard

Case Presentation

The patient was a 78-year-old male with a long history of chronic renal failure. The patient had open heart surgery 3 years ago which was complicated. At that time he experienced the deterioration of his renal function requiring dialysis. An arteriovenous graft (AVG) was inserted at that time. After several months on dialysis, his renal function improved and dialysis was discontinued. Subsequent to that, his AVG thrombosed and since was no longer needed at that time, it was not treated.

Over the past 3 years, the patient's renal function has slowly deteriorated to the point that he again needs to begin dialysis. With no functioning access, a decision was made to evaluate the old abandoned access to determine if it was salvageable. Upon examination the patient was found to have a loop brachial-basilic AVG in his left forearm. The graft was completely collapsed giving a "railroad-track" feel to palpation. Since the graft had been in place only for a short time before it had a spontaneous thrombosis, a decision was made to attempt salvage.

After the patient was prepped and draped, the access was cannulated just above the apex of the loop on the venous side. Because the graft was collapsed, the technique used for cannulation involved inserting the introducer needle with a very flat angle and going all the way through the graft. Then, using index finger as a fulcrum, the tip of the needle was lifted and the needle was slowly retracted. This process was continued until a sudden decrease in tension (a "pop") was felt indicating that the needle tip had entered the space between the two layers of collapsed graft. At that point, an attempt was made to insert a straight hydrophilic guidewire.

Once the guidewire had entered the collapsed graft a short distance, the needle was removed and replaced with a dilator. The dilator was passed up to the tip of the

G.A. Beathard, MD, PhD (✉)
Lifetime Vascular Access, University of Texas Medical Branch,
5135 Holly Terrace, Houston, TX 77056, USA
e-mail: gbeathard@msn.com

© Springer International Publishing AG 2017
A.S. Yevzlin et al. (eds.), *Dialysis Access Cases*,
DOI 10.1007/978-3-319-57500-1_19

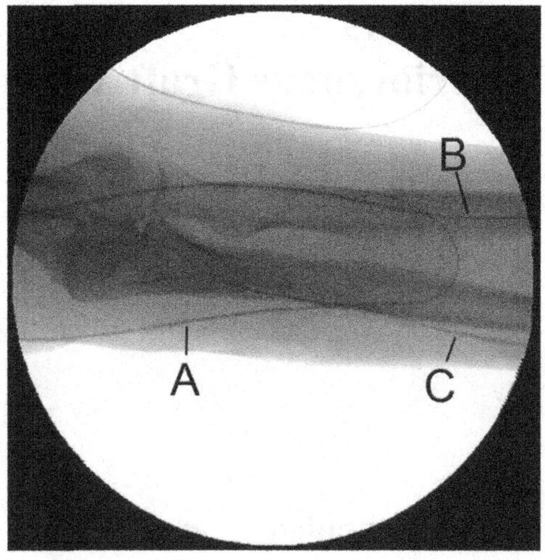

Fig. 19.1 Fluoroscopic view of access with three guidewires in place. *A* Guidewire covering loop of AVG, *B* guidewire covering venous side of AVG, *C* guidewire covering arterial side of AVG

guidewire to stiffen it. This was followed by another attempt to advance the guidewire a short distance. Alternating between advancing the guidewire followed by an advancement of the dilator, the venous anastomosis was eventually crossed. The guidewire was then advanced up to the level of the central veins.

A second cannulation was made in a retrograde direction, just above the apex of the loop on the arterial side. A guidewire was then advanced up and across arterial anastomosis using the same techniques as described above. A third antegrade cannulation was made proximal to the arterial side site, and a guidewire was advanced around the loop of the access to overlap the guidewire which had been inserted at the venous site.

With these three guidewires in place (Fig. 19.1), an 8 × 4 angioplasty balloon was used in a sequential overlapping fashion to open the collapsed graft starting at the venous anastomosis and extending all the way to the arterial anastomosis (Figs. 19.2 and 19.3). At this time, an angiogram of the access revealed blood flow throughout the entire extent of the access (Fig. 19.4).

The newly opened AVG thrombosed overnight. The patient returned to the access center where a standard AVG thrombectomy was performed. This was easily accomplished without complication. The patient went from the access center to dialysis were the AVG was used successfully. At a 3-month follow-up, dialysis was continuing with the access and the patient had encountered no further problems.

Discussion

It is generally agreed that a thrombosed prosthetic bridge graft (AVG) should be treated in a timely manner [1]; however, time definitions are not standardized. The question is how long is too long? Cases that have been occluded for at least 72 h routinely treated successfully, i.e., the case that clots on Friday and presents for

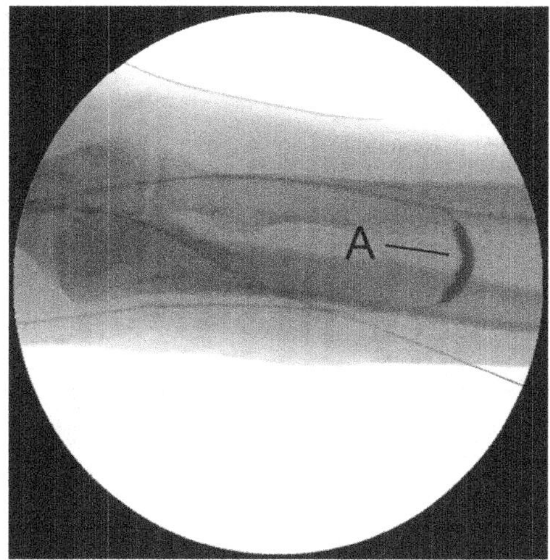

Fig. 19.2 Opening loop of AVG with angioplasty balloon. *A* Angioplasty balloon inflated

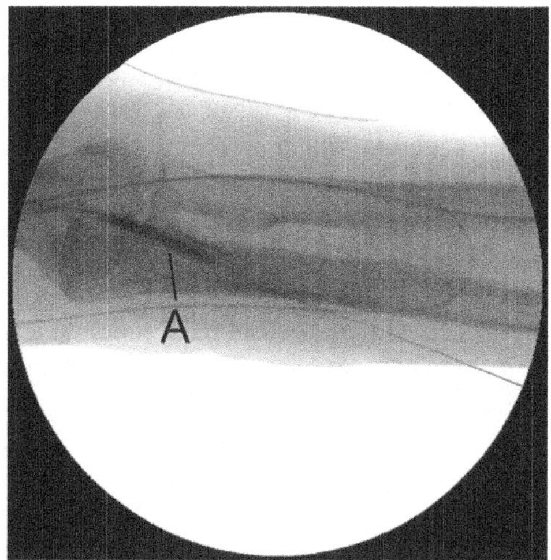

Fig. 19.3 Opening arterial anastomosis with angioplasty balloon. *A* Angioplasty balloon inflated

management on Monday. It is not infrequent that a patient will be referred for a thrombectomy who has been thrombosed for even longer periods. Experience with these cases has taught that although some can be successfully managed, one's chances for success diminishes progressively with time.

As a general rule, the AVG that is completely collapsed, taking on a "railroad-track" appearance, is uniformly associated with salvage failure, so much so that it should not even be attempted. The reason for this relates to the fact that in most instances such a chronically thrombosed graft is the product of chronic problems related to vascular stenosis. Once the access is abandoned, it appears that this process continues to the point of complete obstruction. It is not possible to advance a guidewire across the venous anastomosis, the most common site for stenosis to occur in an AVG. It is of

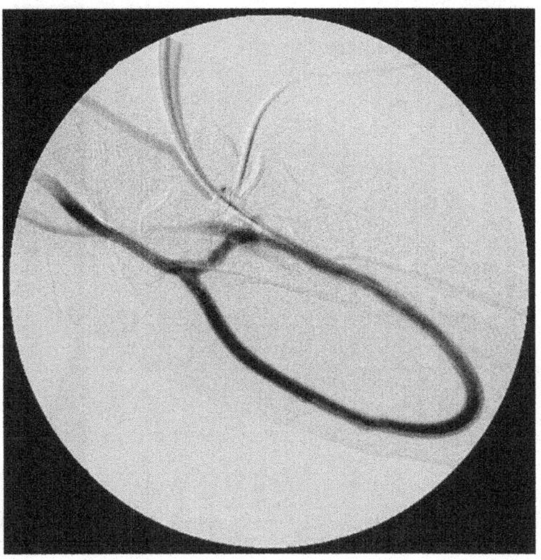

Fig. 19.4 Angiogram of arteriovenous graft performed via vascular catheter passed across arterial anastomosis

interest that even in these cases, it may be possible to cross the arterial anastomosis; however, this is of no value since outflow cannot be restored.

There does appear to be one rare exception to this rule concerning a "dead" AVG. Cases in which a recently created AVG thromboses in association with a return of renal function (recovery of renal function or transplantation) and has been left untreated. These can occasionally (not always) be "resuscitated" even after prolonged periods of occlusion. In one report, 13 cases that met the specific criteria of being placed only a short period of time prior to thrombosis (0.75 to 6 months) associated with a return of renal function and having been left untreated (3 months to 8 years) were attempted over a 12-year period. Salvage was possible in eight (62%) of these cases [2].

References

1. NKF-K/DOQI. Clinical practice guidelines for vascular access. Guideline 21: Treatment of thrombosis and associated stenosis in dialysis av grafts. Am J Kidney Dis. 2001; 37(Suppl 1):S164.
2. Beathard GA. Successful treatment of the chronically thrombosed dialysis access graft: resuscitation of dead grafts. Semin Dial. 2006;19:417–20.

Chapter 20
Arteriovenous Graft Inflow Stenosis

Nabil J. Haddad, Anil K. Agarwal, and Arif Asif

Introduction

Stenosis of arterial inflow reduces blood flow into the access circuit and can cause thrombosis of arteriovenous fistula (AVF) and arteriovenous graft (AVG) as well as failure of maturation of AVF. The inflow segment of AVG comprises of the portion of the feeding artery adjacent to the anastomosis, the artery-graft anastomosis and the first 2 cm of the graft [1]. Hemodynamically significant stenosis of the inflow segment is defined as narrowing of vessel lumen equal to or exceeding 50% of the adjacent normal vascular segment [2]. In this chapter, we discuss an example of inflow stenosis of AVG that was successfully treated with angioplasty.

Case Presentation

A 66-year-old female with ESRD was on hemodialysis via a left forearm arteriovenous graft (AVG). She was noted to have a decline in access flow to less than 300 mL/min with negative arterial pressure exceeding –300 mmHg while using 15-gauge needles for dialysis. She was referred to the access center for evaluation.

N.J. Haddad, MD • A.K. Agarwal, MD (✉)
Division of Nephrology, Department of Internal Medicine, The Ohio State University Wexner Medical Center, Columbus, Ohio, USA
e-mail: nabil.haddad@osumc.edu; anil.agarwal@osumc.edu

A. Asif, MD, MHCM, FASN, FNKF
Department of Medicine, Jersey Shore University Medical Center, Seton Hall-Hackensack-Meridian School of Medicine, Neptune, NJ, USA
e-mail: arifasif@yahoo.com

© Springer International Publishing AG 2017
A.S. Yevzlin et al. (eds.), *Dialysis Access Cases*,
DOI 10.1007/978-3-319-57500-1_20

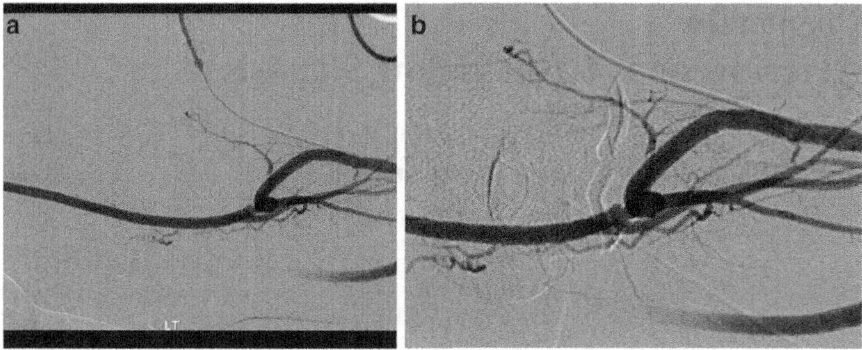

Fig. 20.1 (**a**) Oblique compression-reflux angiogram of the inflow anastomosis of a forearm AVG demonstrates short-segment 50% stenosis at the inflow arterial anastomosis to the brachial artery. (**b**) Slightly different oblique view confirms presence of an approximately 50% anastomotic stenosis

On physical examination, the patient had a left forearm loop AVG with significant accentuation of systolic bruit over the proximal arterial limb in the area of arterial anastomosis. There were no increased pulsations or aneurysmal dilatation of the AVG.

The patient underwent a shuntogram, which showed no significant intragraft stenosis. The venous anastomosis, outflow vein, and central veins did not show stenotic lesion. With manual occlusion of the AVG, a retrograde arteriogram was done which showed about 70% stenosis of the arterial anastomosis (Fig. 20.1a, b). The AVG was then accessed in a retrograde fashion, and a 6-French sheath was placed. Using a 5 × 20-mm angioplasty balloon over a 0.035-inch Glidewire the arterial anastomosis was angioplastied with good results (Fig. 20.2). The patient was dialyzed after the procedure with blood flow of 400 mL/min and adequate venous and arterial pressures.

Discussion

Presence of 50% or more narrowing in the vessel diameter upon angiography usually correlates with a 75% reduction of the vessel cross-sectional area and is considered to be a hemodynamically significant stenosis in many vascular territories. It is well known that AVG has a higher rate of stenosis and thrombosis and requires higher rate of interventions than arteriovenous fistulae (AVF) [3]. Inflow stenosis causes inadequate dialysis and suboptimal flow and can lead to eventual failure of the access.

The frequency of arterial inflow stenosis has traditionally been reported to occur much less frequently in AVG than in AVF. It was reported to account for less than 2%

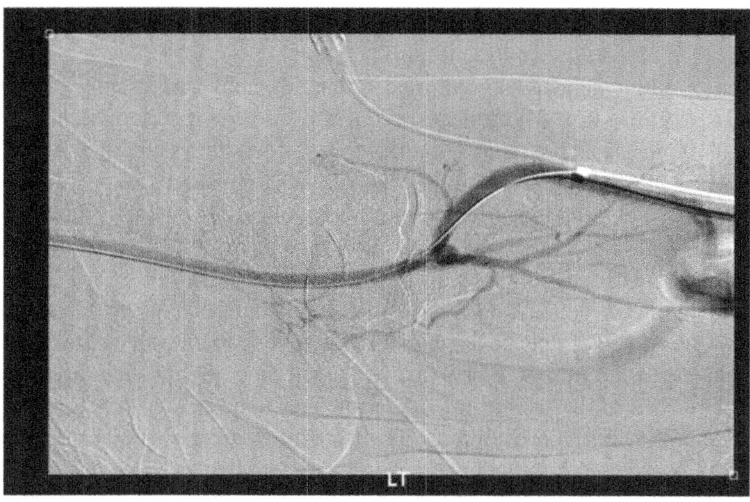

Fig. 20.2 After the angioplasty using a 5×20-mm angioplasty balloon, there is nearly complete disappearance of the angiographically visible stenosis

of stenotic lesions in AVG compared to about 27% of stenotic lesions in AVF, while venous anastomosis and outflow vein stenosis rate was close to 80% of AVG stenotic lesions [4]. More recent studies have shown a higher incidence of arterial inflow stenosis in AVG than previously reported. Using retrograde arteriography, one study reported AVG inflow stenosis rate of 29% versus 40% in AVF [5]. A very similar (25%) incidence of arterial inflow stenosis was noted in another study utilizing antegrade angiography after advancing the angiography catheter in subclavian artery in patients with AVG [6]. Both the studies identified a significant number of these lesions to be in the proximal artery (17% and 23%, respectively). This finding illustrates the importance of a thorough evaluation of the entire inflow, not just the artery-graft anastomosis, especially when the cause of poor inflow is not clear. Differences in the incidence of inflow stenosis among various studies are related primarily to the use of nonstandard definitions and retrospective design of the studies.

Both vascular access monitoring and surveillance are utilized in patients undergoing hemodialysis to detect presence of inadequate access function early so that an intervention can be made in time to avoid loss of access. Physical examination is combined with different surveillance techniques, which include direct access flow monitoring, duplex ultrasound, direct or derived static venous dialysis pressure measurements, and access recirculation calculation [7]. Physical examination may detect up to 80–90% of stenotic lesions in AVF in experienced hands though it may not be as accurate in case of AVG [8]. Signs and symptoms of arterial inflow stenosis include accentuation of systolic bruit over the arterial anastomosis in AVG, decreased access flow, difficult cannulation, inadequate dialysis adequacy, and high negative arterial pressure during hemodialysis (below −200 mm Hg with 15-gauge needles and a blood flow of 400 ml/min) [7].

Arterial inflow stenosis is confirmed by angiography and usually responds to angioplasty [9]. Up to 100% clinical success rate has been reported [6], and patients with AVG stenosis had pre- and post-angioplasty access flows of 588 ± 192 mL/min and 1120 ± 427 mL/min, respectively [5]. With this high success rate, however, it is prudent to keep in mind that arterial anastomosis angioplasty may result in dialysis access-associated steal syndrome in the presence of arterial stenosis distal to the arterial anastomosis [10].

Angiographic visualization of clinically significant 50% or greater stenosis of the inflow anastomosis may be technically difficult due to the angle of takeoff of the graft relative to the inflow artery, due to the presence of overlapping arterial or venous branches in the vicinity of the inflow anastomosis, and due to the overall small size of the arterial anastomosis. There are a few angiographic techniques that could help overcome this challenge and demonstrate the presence of a stenosis in order to help confirm the clinical suspicion of an inflow stenosis and support the decision regarding treatment with balloon angioplasty. These include performing multiple oblique angiographic views of the anastomosis spanning at least a 90-degree arc. These oblique views can be obtained by turning the angiographic tube or by manually rotating the patient's arm or forearm to obtain the proper angles of view. When a high degree of clinical suspicion is present supported by the technique delineated above, retrograde catheterization of the arterial inflow using a small (3-French or 4-French) angiographic catheter inserted retrograde via the graft could help demonstrate the stenosis by allowing direct injection of contrast.

It is possible for a clear angiographic demonstration of an anastomotic stenosis to elude even the most experienced and diligent interventionist, despite multiple angiograms in multiple oblique views. In such circumstances, if the clinical scenario is highly suggestive of an inflow stenosis then a cautious trial of balloon angioplasty using a 4-mm balloon may be performed to help optimize flow and mitigate the risk of access failure. If performed in this empirical manner, in the setting of a strong clinical suspicion of stenosis, then careful observation of the behavior of the balloon and presence of a waist on the balloon during inflation could actually help provide additional information regarding the presence of an anastomotic stenosis. If a waist appears on the balloon and subsequently resolves, it could be a reliable sign that there was some degree of luminal narrowing at the anastomosis. It is important to note that empiric angioplasty of an anastomotic inflow stenosis should be reserved for cases in which there is abundance of clinical evidence suggesting presence of an inflow limiting lesion and should be performed after careful and systematic angiographic evaluation of the inflow vasculature and attempts to optimally demonstrate the anastomosis. In case of recurrence of inflow stenosis after angioplasty, it is possible to use a stent. Some cases may require surgical revision to correct the stenosis.

References

1. Beathard GA, Arnold P, Jackson J, Litchfield T. Aggressive treatment of early fistula failure. Physician operators forum of RMS lifeline. Kidney Int. 2003;64:1487–94.
2. National Kidney Foundation. K/DOQI clinical practice guidelines for vascular access: update 2000. Am J Kidney Dis. 2001;37(Suppl 1):S137–81.
3. Young EW, Dykstra DM, Goodkin DA, Mapes DL, Wolfe RA, Held PJ. Hemodialysis vascular access preferences and outcomes in the Dialysis Outcomes and Practice Patterns Study (DOPPS). Kidney Int. 2002;61(6):2266.
4. Maya ID, Oser R, Saddekni S, Barker J, Allon M. Vascular access stenosis: comparison of arteriovenous grafts and fistulas. Am J Kidney Dis. 2004;44:859–65.
5. Asif A, Gadalean FN, Merrill D, Cherla G, Cepleu CD, Epstein DL, Roth D. Inflow stenosis in arteriovenous fistulas and grafts: a multicenter, prospective study. Kidney Int. 2005;67:1986–92.
6. Khan FA, Vesely TM. Arterial problems associated with dysfunctional hemodialysis grafts: evaluation of patients at high risk for arterial disease. J Vasc Interv Radiol. 2002;13:1109–14.
7. Haddad NJ, Winoto J, Shidham G, Agarwal AK. Hemodialysis access monitoring and surveillance, how and why? Front Biosci. 2012;E4:2396–401.
8. Asif A, Leon C, Orozco-Vargas LC, Krishnamurthy G, Choi KL, Mercado C, Merrill D, Thomas I, Salman L, Artikov S, Bourgoignie JJ. Accuracy of physical examination in the detection of arteriovenous fistula stenosis. Clin J Am Soc Nephrol. 2007;2:1191–4.
9. Brooks JL, Sigley RD, May KJ Jr, Mack RM. Transluminal angioplasty versus *surgical* repair for stenosis of hemodialysis grafts. Am J Surg. 1987;153:530–1.
10. http://www.vascularaccesssociety.com/intro/guidelines. Guideline 07. Treatment of stenosis and thrombosis in AV fistulae and AV grafts. Accessed 1 May 2016.

Chapter 21
Arteriovenous Graft Peri-anastomotic Outflow Stenosis

Tushar J. Vachharajani and Arif Asif

Case Presentation

A 91-year-old patient was receiving hemodialysis via forearm AVG. The patient noticed that it was taking over 30 min for the bleeding to stop after his dialysis needles were withdrawn at the end of each dialysis session. Further review of his dialysis records revealed that the dialysis technician had recorded frequent venous alarms and inability to achieve the prescribed pump speed. The AVG had become increasingly pulsatile and on auscultation had a high-pitched whistling sound at the peri-anastomostic region. The patient was referred for a fistulogram which revealed an outflow stenosis at the peri-anastomotic region, which was successfully angioplastied (Fig. 21.1). The patient was able to receive his regular dialysis following this procedure.

Discussion

Arteriovenous graft (AVG) is a vascular conduit created by using a synthetic material to connect an artery and vein. AVG conduit is used for dialysis when native vessels are unsuitable to create the preferred arteriovenous fistula. The most commonly used synthetic material is polytetrafluoroethylene (PTFE). Several other materials have been used and include polyurethane, cryopreserved vessel, bovine vessel, and bioengineered vein.

T.J. Vachharajani, MD, FASN, FACP (⊠)
W. G. (Bill) Hefner VAMC, 1601 Brenner Ave, Salisbury, NC 28144, USA
e-mail: Tushar.vachharajani@va.gov

A. Asif, MD, MHCM, FASN, FNKF
Department of Medicine, Jersey Shore University Medical Center,
Seton Hall-Hackensack-Meridian School of Medicine, Neptune, NJ, USA

© Springer International Publishing AG 2017
A.S. Yevzlin et al. (eds.), *Dialysis Access Cases*,
DOI 10.1007/978-3-319-57500-1_21

"

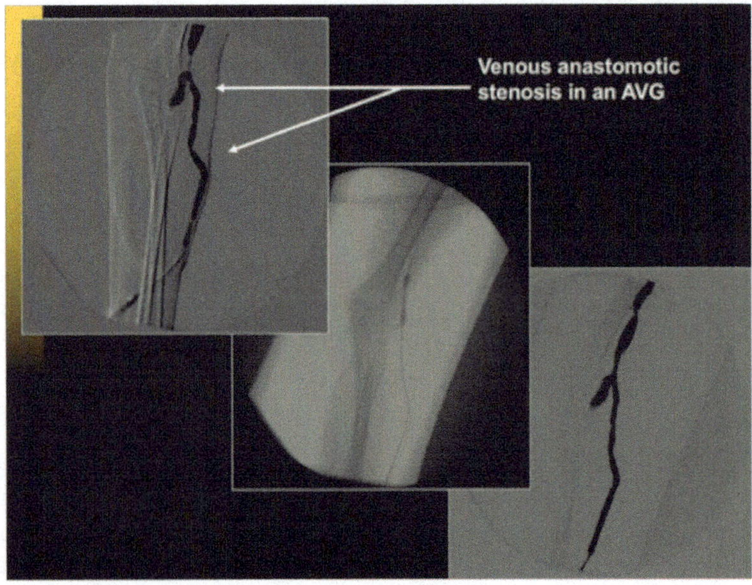

Fig. 21.1 Forearm loop arteriovenous graft. (*a*) Peri-anastomotic stenosis, (*b*) waist on the balloon, (*c*) post-angioplasty image with resolution of stenosis

AVG conduit can be created at various locations in the upper and lower extremity in loop or straight configuration. Exotic locations have been used such as axillary artery to popliteal vein, renal artery to renal vein, necklace AVG, and axillary artery to atrial appendage when all of the common access sites in the upper and lower extremity are exhausted. PTFE graft is generally 6 mm wide for upper extremity conduits and 8 mm wide for lower extremity conduit. The ends can be straight, tapered, or hooded. Generally, AVG can be easily cannulated for dialysis therapy within 2–3 weeks of surgery.

Peri-anastomotic Stenosis

The most common complication encountered with an AVG is the development of venous peri-anastomotic stenosis. The graft-vein anastomotic site stenosis is reported in nearly 60% of dysfunctional AVGs. The progressive worsening of stenosis over time leads to reduction in blood flow eventually leading to AVG thrombosis. The patency of AVG depends on timely diagnosis and intervention of stenosis. Nearly 90% of all thrombotic events in an AVG are associated with an underlying stenosis.

Pathogenesis

The progressive development of neo-intimal hyperplasia at the peri-anastomotic region leads to reduction of luminal diameter increasing resistance to blood flow and eventual thrombosis of AVG. The process of neo-intimal hyperplasia is believed to be related to various factors that include (1) shear stress, (2) turbulence, (3) bio-incompatibility between native vein and PTFE leading to immune-mediated injury, (4) surgical trauma, (5) injury related to frequent needle cannulation during dialysis, and (6) uremic milieu and endothelial dysfunction. The stenotic lesion is characterized on histopathology by increased intimal thickening from smooth muscle proliferation, extracellular matrix, and angiogenesis in the neo-intimal and adventitial layers [1].

Diagnosis

A regular monitoring and surveillance protocol in the dialysis unit can help with the diagnosis of peri-anastomotic stenosis. Monitoring of AVG involves routine physical examination before each dialysis treatment and feeling for a soft pulse and continuous thrill during both systolic and diastolic phases of cardiac cycle. As the degree of stenosis progresses and becomes significant, the thrill diminishes and AVG becomes pulsatile, which is recognized by a bounding pulse. On auscultation with worsening stenosis, the soft continuous bruit becomes a high-pitched sound often described as a "whistling" sound [2] [Multimedia sound]. A thorough physical examination (monitoring) has a high specificity and sensitivity to diagnose peri-anastomotic stenosis. Several surveillance measures such as access blood flow measurements, static venous pressure measurements, or duplex ultrasonography study may be used as additional tools to diagnose peri-anastomotic stenosis.

Along with following the trends in monitoring and surveillance techniques, changes in clinical findings such as prolonged bleeding after withdrawal of dialysis needles (bleeding for more than 15–20 min), poor solute clearance despite compliance with dialysis therapy, and frequent tripping of venous alarms on the dialysis machine are clues that point toward worsening outflow stenosis.

Most definitive method to confirm a peri-anastomotic stenosis is to perform a fistulogram. A stenosis is considered significant on a contrast study if the peri-anastomotic region is narrow by more than 50% compared to the adjacent vein.

Treatment

The treatment options for peri-anastomotic stenosis include endovascular or surgical therapy. The endovascular therapy has distinct advantages since it is minimally invasive, can be performed in an outpatient setting, and helps maintain dialysis therapy schedule. The procedure involves cannulating the access and placing a venous sheath and confirming the presence of significant stenosis on a contrast study. A guide wire is then advanced through the sheath followed by an appropriate sized noncompliant angioplasty balloon. The balloon is inflated over the stenotic segment until the balloon waist is completely effaced. A follow-through contrast study is performed to confirm dilatation of the lesion and absence of any complications.

Outcome

Immediate outcome is almost always successful. The lesion tends to recur over a period of time requiring close monitoring for recurrence and timely intervention to maintain AVG patency. Generally, the lesions tend to recur at 6–9 monthly intervals [3].

References

1. Roy-Chaudhury P, Kelly BS, Miller MA, Reaves A, Armstrong J, Nanayakkara N, Heffelfinger SC. Venous neo-intimal hyperplasia in polytetrafluoroethylene dialysis grafts. Kidney Int. 2001;59(6):2325–34.
2. Beathard G. Physical examination: the forgotten tool. In: Gray JR, Sands JJ, editors. Dialysis access: a multidisciplinary approach. Philadelphia: Lippincott, Williams & Williams; 2002. p. 111–8.
3. Beathard GA. Angioplasty for arteriovenous grafts and fistulae. Semin Nephrol. 2002;22: 202–10.

Chapter 22
Arteriovenous Graft with Traumatic Fistula

Gerald A. Beathard

Case Presentation

The patient was 72-year-old male who had been on dialysis for 6 years using the same forearm loop graft in the left arm. The patient was referred to the dialysis vascular access center because of confusing findings on physical examination detected by the nurse. On physical examination, the patient was found to have a large pseudoaneurysm at both the arterial and at the venous cannulation sites. The arterial side of the graft was pulsatile with a continuous thrill and bruit. Proximal to the pseudoaneurysm at the venous cannulation site, there was no pulse, thrill, or bruit. When a dialysis needle was inserted into this area, there was no blood flow.

Examination of the patient at the access center confirmed the nurse's physical findings. In addition a continuous thrill was present over the pseudoaneurysm at the venous cannulation site, and there was a large vein on the posterior lateral aspect of the forearm, which demonstrated a strong pulse when manually occluded proximally.

After the patient was prepped and draped, the arteriovenous graft (AVG) was cannulated just above the apex of the loop on the venous side. An angiogram was performed which showed the large pseudoaneurysm with blood flow going directly into the venous system via the large vein noted on physical examination. There was no evidence of blood flow above the pseudoaneurysm on the venous side of the AVG (Fig. 22.1). When the angiogram was repeated with the large vein manually occluded, numerous venous collaterals in the forearm were apparent (Fig. 22.2). A third angiogram was performed through a cannulation site on the arterial side of the loop. This showed that this side of the access was patent (Fig. 22.3).

G.A. Beathard, MD, PhD (✉)
Lifetime Vascular Access, University of Texas Medical Branch,
5135 Holly Terrace, Houston, TX 77056, USA
e-mail: gbeathard@msn.com

© Springer International Publishing AG 2017
A.S. Yevzlin et al. (eds.), *Dialysis Access Cases*,
DOI 10.1007/978-3-319-57500-1_22

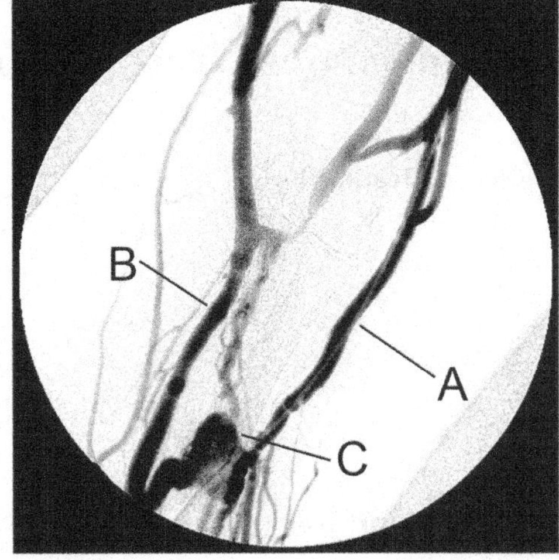

Fig. 22.1 Initial angiogram showing arteriovenous communication between pseudoaneurysm and venous system. *A* Accessory cephalic vein connecting to upper arm cephalic vein, *B* median antebrachial vein connecting to median cubital vein, *C* pseudoaneurysm

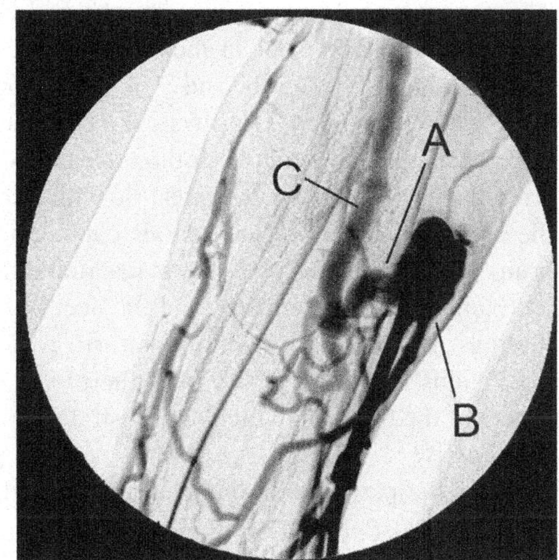

Fig. 22.2 Angiogram with manual occlusion of primary venous outflow of fistula. *A* Fistula connection between pseudoaneurysm and venous system, *B* pseudoaneurysm and graft, *C* vein

Discussion

This case is an example of a traumatic fistula occurring in an AVG. The mechanism by which this occurs is the cannulation needle passing through the wall of the access (in this case the pseudoaneurysm) and entering a vein. When this occurs, if the resistance to flow in the graft is higher than that in the perforated vein, a fistula will develop. Stenosis at the venous anastomosis is a very common occurrence with an AVG; this was undoubtedly a contributing factor to the development of this lesion.

There are two approaches to the treatment of such a fistula. First, the site of the arteriovenous communication can be covered with a stent graft. Prior to this procedure, a thrombectomy would need to be performed on the portion of the AVG above

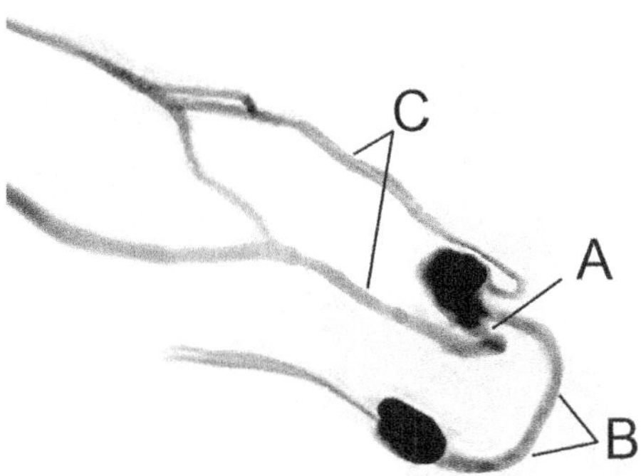

Fig. 22.3 Composite angiogram obtained with arterial side injection. *A* Fistula connection, *B* graft, *C* draining veins

the fistula. Since it is probable that the venous anastomosis is stenotic, it would also need to be treated. Second, the patient could be referred to surgery for revision. Since the patient has two large pseudoaneurysms, these could be addressed at the same time.

In a case such as this, one should not ignore the fact that the veins, which have been involved in the fistula, could very well be excellent candidates for the creation of a secondary fistula.

Chapter 23
Pseudoaneurysm Treated with Stent Graft

Gerald A. Beathard

Case Presentation

The patient was a 48-year-old male who had been on dialysis for 6 years who was referred to the vascular access center because of poor flow in his left upper arm loop arteriovenous graft (AVG). Examination revealed that the patient had a large pseudoaneurysm at the venous cannulation site on the AVG. The AVG was hyperpulsatile, and there was a marked thrill present over the venous anastomosis. After the patient was prepped and draped, an angiogram was performed which showed marked stenosis of the venous anastomosis. This was successfully treated with angioplasty. The angiogram also revealed a large irregularly shaped pseudoaneurysm (Fig. 23.1). It was decided to treat this defect with a stent-graft. A 10 × 8 stent-graft was inserted through a sheath and positioned across the pseudoaneurysm (Fig. 23.2). After it was deployed, an 8 × 4 angioplasty balloon was inflated within the stent-graft to assure good apposition with the wall of the AVG (Fig. 23.3). And angiogram performed after the procedures showed that the pseudoaneurysm had been excluded from the lumen and that there was good flow throughout the AVG (Fig. 23.4). In order to collapse the pseudoaneurysm, the blood and old thrombus within it was then evacuated using a large gage needle attached to a syringe.

G.A. Beathard, MD, PhD (✉)
Lifetime Vascular Access, University of Texas Medical Branch,
5135 Holly Terrace, Houston, TX 77056, USA
e-mail: gbeathard@msn.com

© Springer International Publishing AG 2017
A.S. Yevzlin et al. (eds.), *Dialysis Access Cases*,
DOI 10.1007/978-3-319-57500-1_23

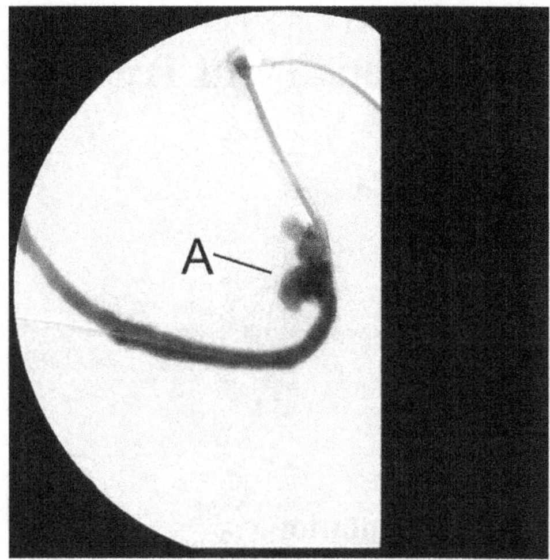

Fig. 23.1 Angiogram of arteriovenous graft. *A* Pseudoaneurysm

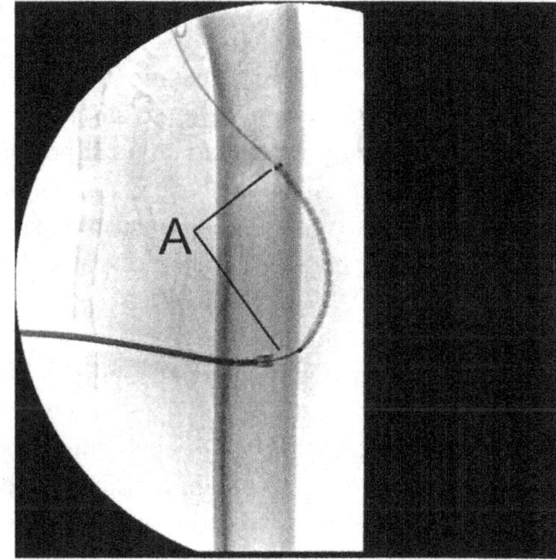

Fig. 23.2 Positioning of stent across the mouth of pseudoaneurysm. *A* Stent graft

Discussion

Since the development of pseudoaneurysm is a frequent event in patients receiving dialysis with an AVG, decisions related to their management are an important aspect of the medical practice related to dialysis vascular access. There is general agreement that a pseudoaneurysm is treated most effectively by surgery. However, in recent years there have been a number of reports of cases treated with stent-grafts [1–7]. Although there is a serious question as to the cost-effectiveness of a stent-graft versus surgery, this can be an effective approach to the problem. There are drawbacks, however. The treatment of a pseudoaneurysm with a stent-graft is not always successful [7]. The pseudoaneurysm is generally associated with increased intravascular pressure due to a

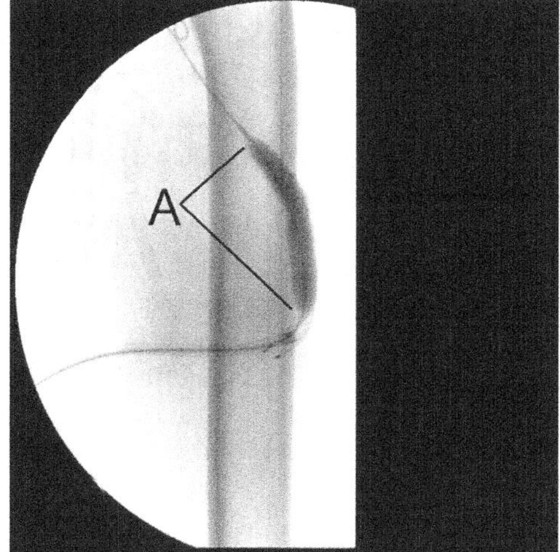

Fig. 23.3 Angioplasty balloon being used to seat stent graft and assure good wall apposition. *A* Angioplasty balloon

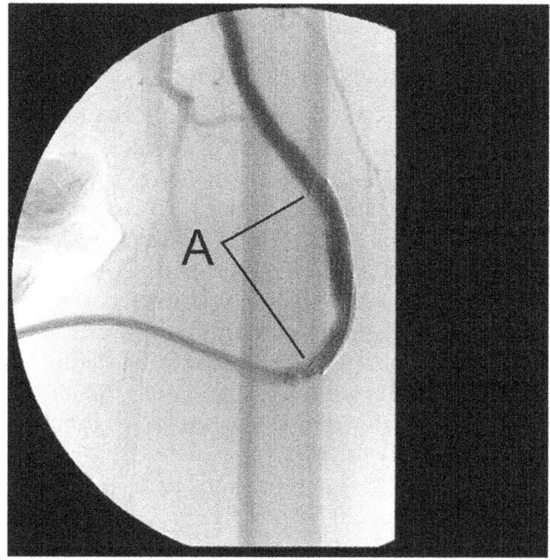

Fig. 23.4 Post-procedure angiogram showing stent graft in place with no evidence of filling of pseudoaneurysm and good flow throughout the arteriovenous graft

downstream stenosis. Prior to the treatment of the pseudoaneurysm, the access outflow should be evaluated for stenosis, and all stenotic lesions should be treated. It has been recommended that very careful angiographic visualization be made in order to determine the size of the mouth of the pseudoaneurysm [8]. A pseudoaneurysm with a large mouth is difficult to treat and may predispose to herniation of the stent-graft into the deformity. It is also important to use a stent-graft that is long enough so that there will be approximately a 1 - 2 cm normal graft landing zone on either side of the pseudoaneurysm. Failure to do this can lead to leakage around the stent-graft and refilling of the pseudoaneurysm.

Prior to beginning the procedure, a careful physical examination of the access should be made to exclude the presence of skin necrosis or an infection, both of which should be considered a contraindication to the procedure.

When used for this purpose, the stent graft is frequently placed within the cannulation zone of the AVG. While cannulation can be performed successfully through the device, it can predispose to complications [7, 9, 10]. It has been recommended by some that if a stent-graft is used to treat a pseudoaneurysm, it not be cannulated [7, 10].

Two major complications have been reported in association with treatment of pseudoaneurysms with stent-grafts – infection and perforation of the skin by a fractured strut of the stent-graft. Both of these are felt to be related to cannulating through the stent-graft. Cannulation of the AVG through the stent-graft eventually results in fracture of some of the struts that compose its nitinol framework [7]. It is possible for the fractured strut to perforate the skin of the patient and, in addition to providing an obvious route for infection into the AVG, present the opportunity for injury to the dialysis staff as they cleanse and cannulate the access [10].

References

1. Sapoval MR, Turmel-Rodrigues LA, Raynaud AC, Bourquelot P, Rodrigue H, Gaux JC. Cragg covered stents in hemodialysis access: initial and midterm results. J Vasc Interv Radiol. 1996;7(3):335–42.
2. Hausegger KA, Tiessenhausen K, Klimpfinger M, Raith J, Hauser H, Tauss J. Aneurysms of hemodialysis access grafts: treatment with covered stents: a report of three cases. Cardiovasc Intervent Radiol. 1998;21(4):334–7.
3. Rabindranauth P, Shindelman L. Transluminal stent-graft repair for pseudoaneurysm of PTFE hemodialysis grafts. J Endovasc Surg. 1998;5(2):138–41.
4. Najibi S, Bush RL, Terramani TT, Chaikof EL, Gunnoud AB, Lumsden AB, et al. Covered stent exclusion of dialysis access pseudoaneurysms. J Surg Res. 2002;106(1):15–9.
5. Silas AM, Bettmann MA. Utility of covered stents for revision of aging failing synthetic hemodialysis grafts: a report of three cases. Cardiovasc Intervent Radiol. 2003;26(6):550–3.
6. Ryan JM, Dumbleton SA, Doherty J, Smith TP. Technical innovation. Using a covered stent (wallgraft) to treat pseudoaneurysms of dialysis grafts and fistulas. AJR Am J Roentgenol. 2003;180(4):1067–71.
7. Vesely TM. Use of stent grafts to repair hemodialysis graft-related pseudoaneurysms. J Vasc Interv Radiol. 2005;16(10):1301–7.
8. Florescu MC, Fillaus JA, Plumb TJ. How I do it: endovascular treatment of arteriovenous graft pseudoaneurysms-watch out for the mouth. Semin Dial. 2014;27(2):205–9.
9. Gadalean F. Covered stent outcomes for arteriovenous hemodialysis access salvage. J Am Soc Nephrol. 2008;19:899A.
10. Asif A, Gadalean F, Eid N, Merrill D, Salman L. Stent graft infection and protrusion through the skin: clinical considerations and potential medico-legal ramifications. Semin Dial. 2010;23(5):540–2.

Chapter 24
Superior Vena Caval Stenosis

Gerald A. Beathard

Case Presentation

The patient was a 52-year-old male who has been on dialysis for 4 months using a right internal jugular tunneled dialysis catheter. The patient had a loop graft placed in his right forearm 10 days ago. He was referred to the dialysis access center because of swelling of the right arm.

Physical examination revealed the presence of a newly placed loop graft in the right forearm. The right arm was markedly swollen. There was evidence of old catheter placements on both sides of the patient's thorax.

After the right arm was prepped and draped, an angiogram was performed (Fig. 24.1, arrow). This showed that all veins were patent up to the level of the lower superior vena cava. An area of severe stenosis was noted at that point. It was possible to pass a guidewire through this lesion down into the inferior vena cava. Angioplasty was performed using a 14 × 4 angioplasty balloon with good result (Fig. 24.2, arrow). A post-angioplasty angiogram showed good flow through the superior vena cava (Fig. 24.3). In addition, the azygos vein which has not been visualized earlier was apparent and was greatly enlarged (Figs. 24.3 and 24.4, arrow).

By the following day, the patient's arm swelling had resolved, and the patient was able to dialyze without problems.

G.A. Beathard, MD, PhD (✉)
Lifetime Vascular Access, University of Texas Medical Branch,
5135 Holly Terrace, Houston, TX 77056, USA
e-mail: gbeathard@msn.com

© Springer International Publishing AG 2017
A.S. Yevzlin et al. (eds.), *Dialysis Access Cases*,
DOI 10.1007/978-3-319-57500-1_24

Fig. 24.1 Initial
angiogram showing
stenosis of superior vena
cava (*arrow*)

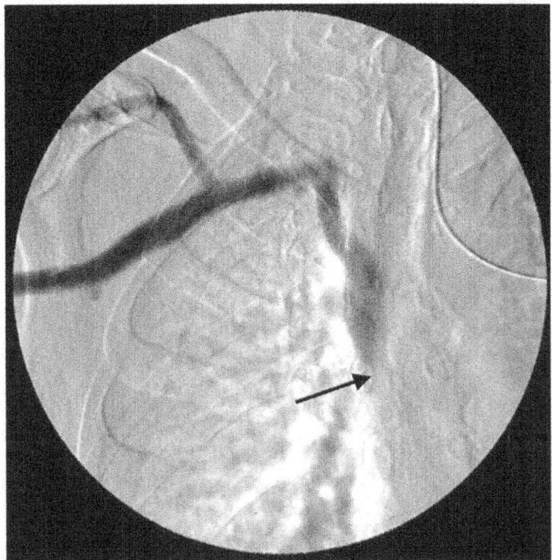

Fig. 24.2 Angioplasty of
stenotic lesion showing full
effacement of angioplasty
balloon (*arrow*)

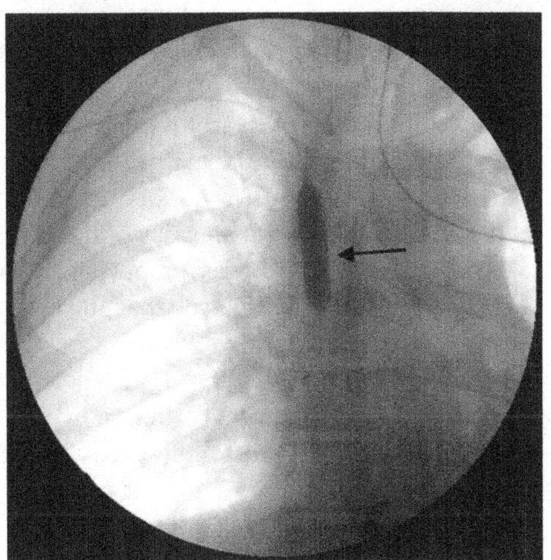

Discussion

This represents a typical case in which previous tunneled dialysis catheters have
had resulted in central venous stenosis, which became evident once the peripheral
access is created.

Although the problem can develop without an identifiable antecedent [1], central
vein stenosis is generally the result of an identifiable instance in which a foreign
indwelling object was introduced. The most common culprit is the central venous
catheter. In general, the frequency of central venous stenosis is directly proportional
to the frequency of dialysis catheter usage. The location of the central venous lesion

Fig. 24.3 Post-procedure angiogram showing good flow in superior vena cava. *Arrow* indicates the opening of the azygos vein seen on end

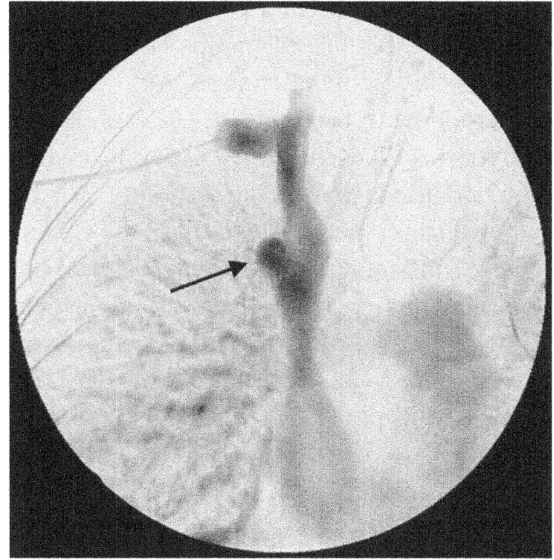

Fig. 24.4 Appearance of late frames of angiogram showing azygos vein (*arrows*)

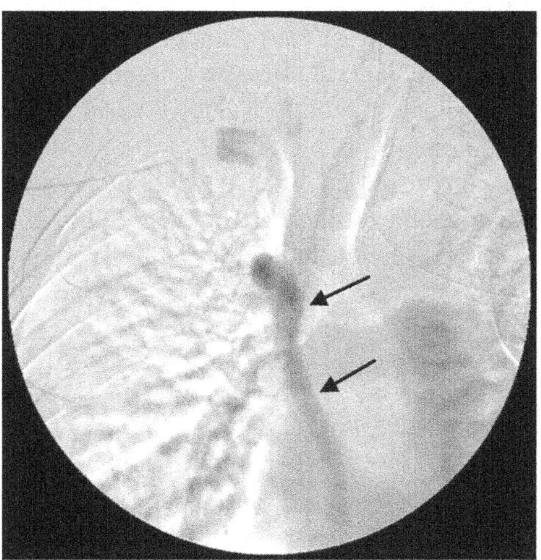

is variable depending somewhat upon the vein into which the catheter was introduced. With internal jugular placement, the brachiocephalic is the vein most commonly affected.

In the dialysis patient, central venous flow and pressure is markedly increased due to the diversion of arterial flow into the venous circuit. Central venous obstruction or even non-obstructive stenosis makes the patient particularly vulnerable to the development of severe, progressive edema of the ipsilateral arm, shoulder, breast, neck, and face [2–5]. Pleural effusion and even a superior vena caval syndrome can be produced. These symptoms cause considerable patient discomfort and place the patient at risk for serious complications. Dialysis therapy becomes increasing difficult due to problems with cannulation.

It is of interest that a peripheral lesion even as far proximal as the axillary vein rarely causes edema of the ipsilateral extremity while just a few centimeters more central, a lesion of the same severity will cause massive edema. In fact, this symptom is considered to be diagnostic of central vein stenosis. The most probable explanation for this relates to the fact that the central vein is the final common pathway for all blood drainage from the extremity. When it is obstructed, the increased resistance to flow that is created affects all of the venous structures in the extremity. This global increase in venous pressure causes the exudation of fluid into the tissue. Stenosis anywhere more peripheral causes only a local effect; other veins provide a free outlet allowing the extremity to avoid a generalized increase in venous pressure.

Angioplasty has become the standard of care as the first-line treatment of these lesions. Multiple studies have demonstrated a variable technical success rate ranging from 70% to 90% [6–13]. The primary and cumulative patency rates for PTA treatment of CVS have demonstrated wide variability. The reported 6-month primary patency rate has ranged from 23% to 63% with a cumulative patency rate range of 29–100%. The 12-month primary patency rate range has been 12–50% with a cumulative patency rate range of 13–100% [6–13].

When there is obstruction to blood flow in the central venous system, the major route for collateral flow is retrograde flow through any venous channels connected to the superior vena cava . However, in this instance, flow in the system is retrograde routing the venous drainage of the thoracic central venous system. The retrograde flow directs blood either beyond the point of obstruction to the superior vena cava or carries the throacic central venous blood to the inferior vena cava. This later situation was obviously the case in this patient judging by the marked enlargement of the azygos vein that was noted on the post-procedure angiograms.

References

1. Oguzkurt L, Tercan F, Yildirim S, Torun D. Central venous stenosis in haemodialysis patients without a previous history of catheter placement. Eur J Radiol. 2005;55:237.
2. Korzets A, Chagnac A, Ori Y, et al. Subclavian vein stenosis, permanent cardiac pacemakers and the haemodialysed patient. Nephron. 1991;58:103.
3. Teruya TH, Abou-Zamzam AM Jr, Limm W, Wong L. Symptomatic subclavian vein stenosis and occlusion in hemodialysis patients with transvenous pacemakers. Ann Vasc Surg. 2003;17:526.
4. Asif A, Salman L, Carrillo RG, et al. Patency rates for angioplasty in the treatment of pacemaker-induced central venous stenosis in hemodialysis patients: results of a multi-center study. Semin Dial. 2009;22:671.
5. Sticherling C, Chough SP, Baker RL, et al. Prevalence of central venous occlusion in patients with chronic defibrillator leads. Am Heart J. 2001;141:813.
6. Glanz S, Gordon DH, Lipkowitz GS, et al. Axillary and subclavian vein stenosis: percutaneous angioplasty. Radiology. 1988;168:371.
7. Beathard GA. Percutaneous transvenous angioplasty in the treatment of vascular access stenosis. Kidney Int. 1992;42:1390.
8. Beathard GA. The treatment of vascular access graft dysfunction: a nephrologist's view and experience. Adv Ren Replace Ther. 1994;1:131.
9. Kovalik EC, Newman GE, Suhocki P, et al. Correction of central venous stenoses: use of angioplasty and vascular Wallstents. Kidney Int. 1994;45:1177.

10. Quinn SF, Schuman ES, Demlow TA, et al. Percutaneous transluminal angioplasty versus endovascular stent placement in the treatment of venous stenoses in patients undergoing hemodialysis: intermediate results. J Vasc Interv Radiol. 1995;6:851.
11. Dammers R, de Haan MW, Planken NR, et al. Central vein obstruction in hemodialysis patients: results of radiological and surgical intervention. Eur J Vasc Endovasc Surg. 2003;26:317.
12. Surowiec SM, Fegley AJ, Tanski WJ, et al. Endovascular management of central venous stenoses in the hemodialysis patient: results of percutaneous therapy. Vasc Endovasc Surg. 2004;38:349.
13. Bakken AM, Protack CD, Saad WE, et al. Long-term outcomes of primary angioplasty and primary stenting of central venous stenosis in hemodialysis patients. J Vasc Surg. 2007;45:776.

Chapter 25
Thrombosed Infected Graft

Gerald A. Beathard

Case Presentation

The patient was a 38-year-old female who had been on dialysis for 6 months with a left forearm loop graft who was referred to the dialysis access clinic because of thrombosis. After the patient was prepped and draped, the graft was cannulated using an 18 ga introducer needle slightly above the apex of the loop on the venous side of the graft. A guidewire was introduced, and the needle was removed. At this time, purulent material began to exude from the cannulation site (Fig. 25.1). The procedure was discontinued, and the patient was referred to the hospital for management. Blood cultures performed at the hospital were negative; however, purulent material draining from the graft was positive for *Staphylococcus epidermidis*.

Discussion

Infection of the AVG is a serious complication. It has been reported to account for 20% of all dialysis access complications [1] and to be the second leading cause of graft loss. Additionally it is the second most common cause of death in this group of patients [2–5]. In one report, infection was reported to occur at a frequency of 1.3 episodes per 100 dialysis months and be associated with bacteremia at a rate of 0.7 cases per 100 dialysis months [6]. In a prospective Canadian study in which surveillance for hemodialysis-related bloodstream infections was performed in 11 centers during a 6-month period, it was found that the relative risk for bloodstream infection

G.A. Beathard, MD, PhD (✉)
Lifetime Vascular Access, University of Texas Medical Branch,
5135 Holly Terrace, Houston, TX 77056, USA
e-mail: gbeathard@msn.com

© Springer International Publishing AG 2017
A.S. Yevzlin et al. (eds.), *Dialysis Access Cases*,
DOI 10.1007/978-3-319-57500-1_25

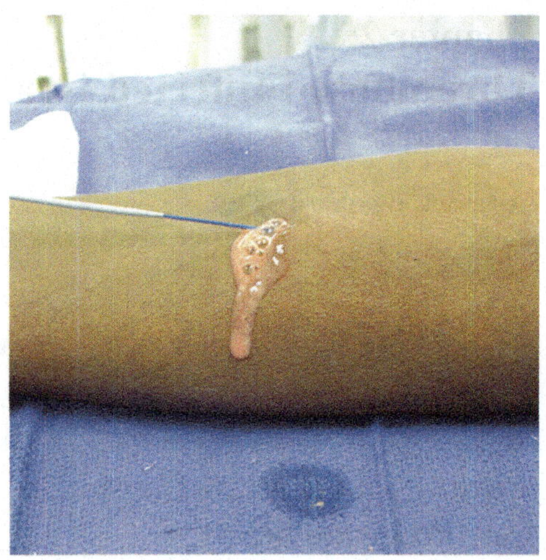

Fig. 25.1 Appearance of cannulation site draining purulent material

with an AVG access was 2.5 per 1000 dialysis procedures. This was in comparison to a rate of 0.2 in patients receiving dialysis via an arteriovenous fistula (AVF) [7].

The diagnosis of an infected graft is usually evident on physical exam, but it is not uncommon for these patients to present with atypical symptoms and a lack of clinical findings. In some instances, the first indication of an infection is when the graft is cannulated and puss is extruded. In these cases, the graft is generally thrombosed, and the clinical picture elicited is an indication that the infection is actually within the graft lumen rather than around it.

An interesting aspect of this case is the question - what would have happened if the thrombosed graft had been abandoned rather than making an attempt to salvage? The patient was asymptomatic, and there was no external evidence of graft infection, yet the lumen was filled with purulent material. There are reports of dialysis patients presenting with fever of unknown origin being found to have evidence of infection in an old abandoned graft [8–13]. It is reasonable to postulate that had this graft been abandoned, this patient might eventually fall into this category.

References

1. Butterly DW. A quality improvement program for hemodialysis vascular access. Adv Ren Replace Ther. 1994;1(2):163–6.
2. Sarnak MJ, Jaber BL. Mortality caused by sepsis in patients with end-stage renal disease compared with the general population. Kidney Int. 2000;58(4):1758–64.
3. Butterly DW, Schwab SJ. Dialysis access infections. Curr Opin Nephrol Hypertens. 2000;9(6):631–5.
4. Lafrance JP, Rahme E, Lelorier J, Iqbal S. Vascular access-related infections: definitions, incidence rates, and risk factors. Am J Kidney Dis. 2008;52(5):982–93.
5. Sexton DJ. Vascular access infections in patients undergoing dialysis with special emphasis on the role and treatment of *Staphylococcus aureus*. Infect Dis Clin N Am. 2001;15(3):731–42, vii

  6. Kaplowitz LG, Comstock JA, Landwehr DM, Dalton HP, Mayhall CG. A prospective study of infections in hemodialysis patients: patient hygiene and other risk factors for infection. Infect Control Hosp Epidemiol. 1988;9(12):534–41.
  7. Taylor G, Gravel D, Johnston L, Embil J, Holton D, Paton S. Prospective surveillance for primary bloodstream infections occurring in Canadian hemodialysis units. Infect Control Hosp Epidemiol. 2002;23(12):716–20.
  8. Ayus JC, Sheikh-Hamad D. Silent infection in clotted hemodialysis access grafts. J Am Soc Nephrol. 1998;9(7):1314–7.
  9. Sheikh-Hamad D, Ayus JC. The patient with a clotted PTFE graft developing fever. Nephrol Dial Transplant. 1998;13(9):2392–3.
 10. Fishbane S, Maesaka J. Occult nonfunctioning AV graft infection as a cause of refractory anemia and low serum albumin in hemodialysis patients [abstract]. J Am Soc Nephrol. 1999;10:278A–9A.
 11. Nassar GM, Ayus JC. Clotted arteriovenous grafts: a silent source of infection. Semin Dial. 2000;13(1):1–3.
 12. Nassar GM, Ayus JC. Infectious complications of the hemodialysis access. Kidney Int. 2001;60(1):1–13.
 13. Nassar GM, Fishbane S, Ayus JC. Occult infection of old nonfunctioning arteriovenous grafts: a novel cause of erythropoietin resistance and chronic inflammation in hemodialysis patients. Kidney Int Suppl. 2002;80:49–54.

Chapter 26
Arteriovenous Thigh Graft Cannulated from Contralateral Side

Gerald A. Beathard

Case Presentation

The patient was a 29-year-old female with a left thigh graft referred to the dialysis access center because of low blood flow. The patient had been on dialysis for 8 years and had multiple catheters and prior arteriovenous accesses. She was known to have bilateral jugular and subclavian vein occlusions. Examination of the access revealed that it was hyper-pulsatile with a high-pitched bruit present over the venous anastomosis

After the patient was prepped and draped, the arteriovenous graft (AVG) was cannulated, and an angiogram was performed (Fig. 26.1). This showed that the patient had severe stenosis at the venous anastomosis. Multiple attempts to pass a guidewire across anastomosis were unsuccessful. While attempting to pass the guidewire, the access thrombosed. At this time, the right femoral vein was cannulated, a sheath was inserted, and a guidewire was introduced. Using a Kumpe vascular catheter, the guidewire was directed into the left common iliac vein at the level of the bifurcation of the inferior vena cava. The guidewire was then advanced down the iliac vein (Fig. 26.2). Using the vascular catheter, it was possible to cross the venous anastomosis of the AVG. An 8 × 4 angioplasty balloon was introduced over the guidewire and advance down to the stenotic lesion, which was dilated (Fig. 26.3). A post-angioplasty angiogram showed a good result (Fig. 26.4).

G.A. Beathard, MD, PhD (✉)
Lifetime Vascular Access, University of Texas Medical Branch,
5135 Holly Terrace, Houston, TX 77056, USA
e-mail: gbeathard@msn.com

© Springer International Publishing AG 2017
A.S. Yevzlin et al. (eds.), *Dialysis Access Cases*,
DOI 10.1007/978-3-319-57500-1_26

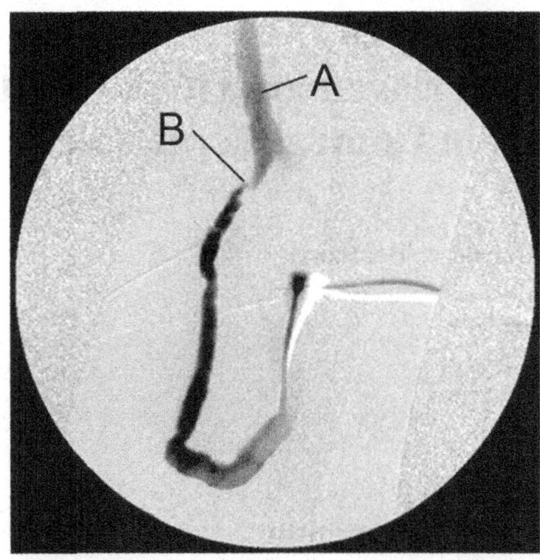

Fig. 26.1 Initial angiogram. *A* External iliac vein, *B* venous anastomosis showing severe stenosis

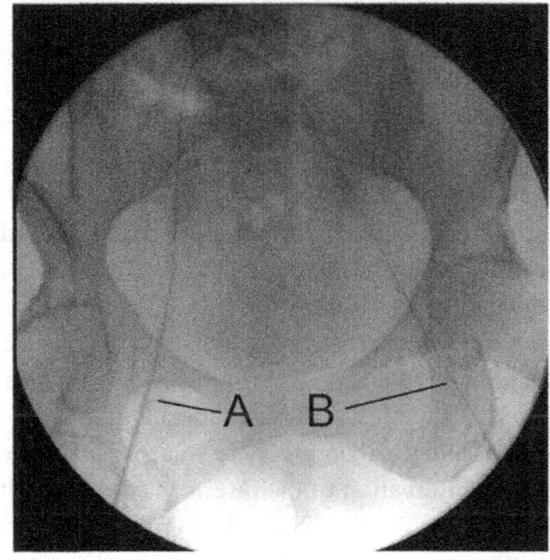

Fig. 26.2 Course of guidewire. *A* Guidewire passing the iliac of *right side*, *B* guidewire extending down through the iliac of left side and across venous anastomosis

Discussion

In AVG cases with severe stenosis of the venous anastomosis, it may be very difficult to pass a guidewire across the site. Frequently, the use of a vascular catheter to stiffen the wire and direct its tip leads to success. However, there are instances in which even after multiple attempts, the stenotic lesion cannot be crossed. In these cases, advancing the guidewire from the opposite direction may solve the problem.

Fig. 26.4 Post-angioplasty angiogram. *A* Site of lesion

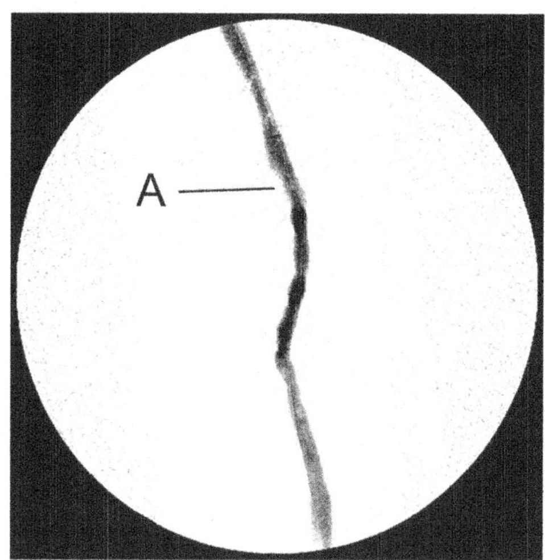

Fig. 26.3 Angioplasty of venous anastomosis. *A* Site of lesion

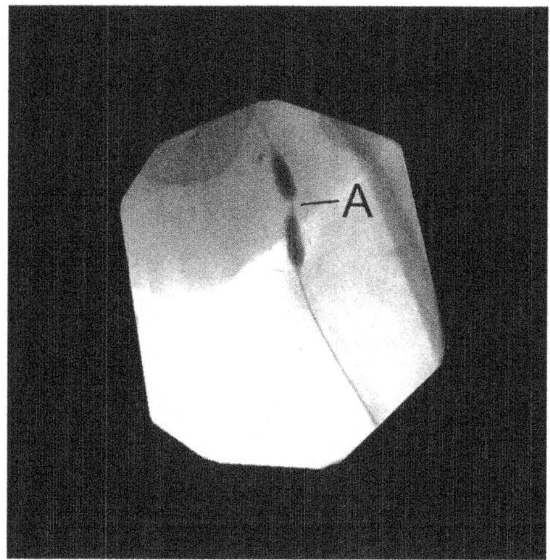

Part III
Dialysis Catheter

Chapter 27
Central Vein Stenosis with Angioplasty Prior to Catheter Placement

Gerald A. Beathard

Case Presentation

The patient was a 54-year-old female who had been on dialysis for proximally 6 years. During this period of time, the patient has had several central venous dialysis catheters and several peripheral arteriovenous accesses. The patient was referred to the vascular access facility for the placement of a cuffed tunneled catheter to use as a bridge until her newly placed left brachial-basilic transposition fistula matured.

Physical examination revealed several old peripheral arteriovenous accesses which had been abandoned, as well as a newly placed left upper arm fistula. The patient had several scars in the subclavicular area bilaterally indicating sites of previous cuffed tunneled catheters. Ultrasound examination of the patient's right internal jugular vein revealed that it was patent. However, it did not dilate significantly with a Valsalva maneuver.

It was decided to cannulate the right internal jugular vein and evaluate with an angiogram to determine if it would be usable for catheter placement. The vein was cannulated with a micropuncture needle using ultrasound guidance, and an angiogram was performed. This study (Fig. 27.1) showed that although the vein was patent, there was marked stenosis of the central portion of the vein with the presence of multiple collaterals. A 0.35 inch guidewire was inserted and passed into the inferior vena cava. A sheath was inserted at the cannulation site, and the internal jugular vein was dilated using a 10 mm × 4 cm angioplasty balloon (Fig. 27.2). After the dilatation, the stenotic area was significantly improved (Fig. 27.3). A 28 cm cuffed tunneled dialysis catheter was then inserted (Fig. 27.4).

The catheter functioned well when tested using a 10 mL syringe. The patient was sent to the dialysis facility where dialysis was performed without difficulty.

G.A. Beathard, MD, PhD (✉)
Lifetime Vascular Access, University of Texas Medical Branch,
5135 Holly Terrace, Houston, TX 77056, USA
e-mail: gbeathard@msn.com

© Springer International Publishing AG 2017
A.S. Yevzlin et al. (eds.), *Dialysis Access Cases*,
DOI 10.1007/978-3-319-57500-1_27

Fig. 27.1 Initial
angiogram showing
stenosis of internal jugular
vein. *A* Cannulation site, *B*
stenotic area

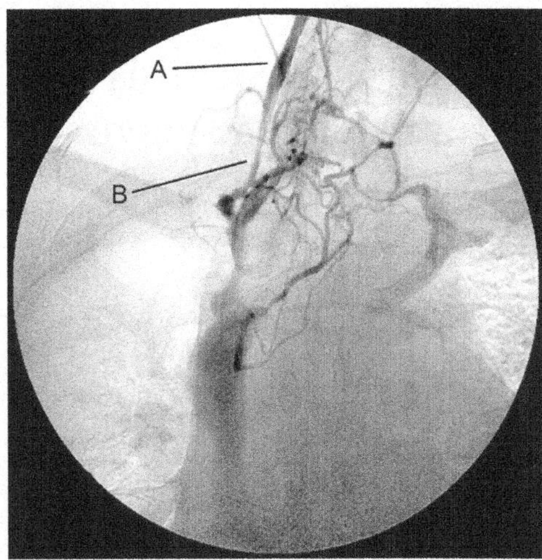

Fig. 27.2 Angioplasty of
stenotic zone

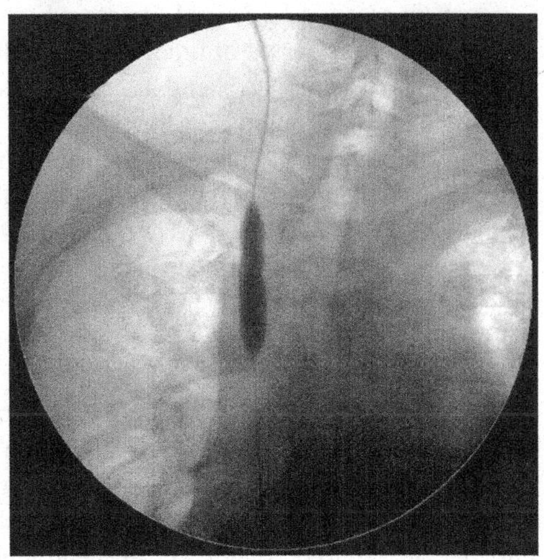

Discussion

Stenosis of the internal jugular vein among today's dialysis population is not an unusual finding. The challenge is in finding a site suitable for catheter insertion when the need arises. In this case, a problem with the vein was suspected at the time it was evaluated with ultrasound because of failure of the vein to dilate significantly with a Valsalva maneuver. In these cases, if the vein can be cannulated it is worthwhile to do so in order to completely evaluate the anatomy. Some of these are salvageable as was the case in this patient. If it is possible to pass a guidewire, an angioplasty balloon can be used to treat the stenosis. In this case, cannulation of the vein at the optimal catheter insertion site was possible since the problem was lower

Fig. 27.3 Post-angioplasty angiogram. *A* Previously stenotic zone

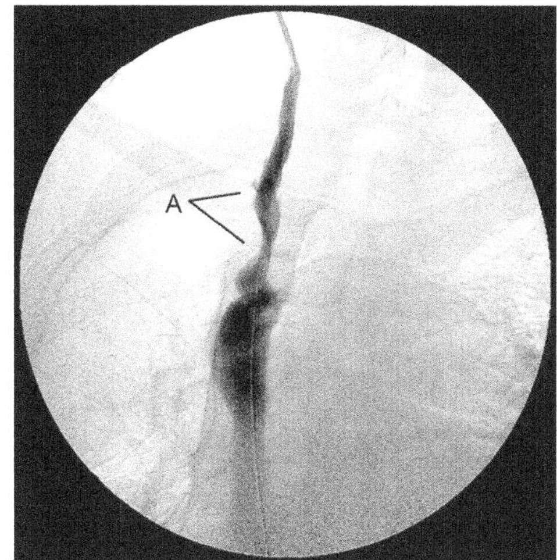

Fig. 27.4 Tunneled catheter in place

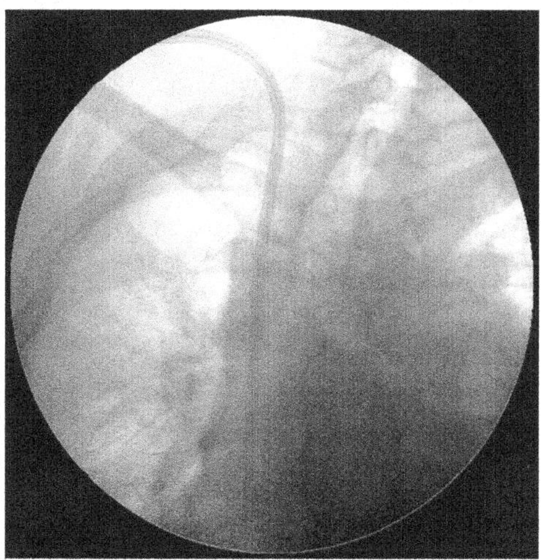

down. Once this area had been dilated, insertion of the catheter was easily accomplished.

In a case such as this, removal of the catheter at a later date may be followed by a loss of the access site since the process of sclerosis was already far advanced at the time of placement and will very likely progress. In a patient who is totally catheter dependent, this loss would be very problematic. In such a case, a plan for continued dialysis access should be made in advance. Such a patient might be an excellent candidate for a HeRO device as long as the venous access is available.

Chapter 28
Catheter Placement in a Stenotic Central Vein

Alexander S. Yevzlin

Case Presentation

A 54-year-old male was referred for hemodialysis initiation in the setting of multiple myeloma requiring chemotherapy.

After informed consent was obtained, the patient was prepped and draped in the usual sterile fashion overlying the right side of the neck and chest, and 15 mL of lidocaine was administered in the area overlying the right side of the neck and chest. Ultrasound guidance was used to localize the right internal jugular vein. Cannulation was performed on the RIJ vein with a 4-French introducer needle. A wire was inserted into the needle under fluoroscopic guidance but could not be passed into the right atrium. An introducer was then passed over the wire and the wire was removed.

A central venogram was performed revealing a severe SVC stenosis (Fig. 28.1). The 4-French introducer was hence changed to a 6-French introducer, and a hydrophilic guidewire and guide catheter were used to cross the lesion. An 8 mm × 4 cm angioplasty balloon was introduced over a wire. Balloon angioplasty was performed applying 15 atm pressure for 20 s. Post-angioplasty venogram revealed stenosis <50% (Fig. 28.2).

Subsequently a 1-cm incision was performed at the insertion point of the dilator as well as below the clavicle, and a tunnel was created using blunt dissection from the two incision points. Next, a 23-cm cuffed dialysis catheter was inserted into the lumen of the tunnel. Under fluoroscopic guidance, a dilator-introducer combination was then inserted into the central vasculature. The dilator and wire were then removed, and the catheter was inserted into the introducer. The introducer was

A.S. Yevzlin, MD, FASN (✉)
University of Michigan, Ann Arbor, MI, USA
e-mail: asy@medicine.wisc.edu

© Springer International Publishing AG 2017
A.S. Yevzlin et al. (eds.), *Dialysis Access Cases*,
DOI 10.1007/978-3-319-57500-1_28

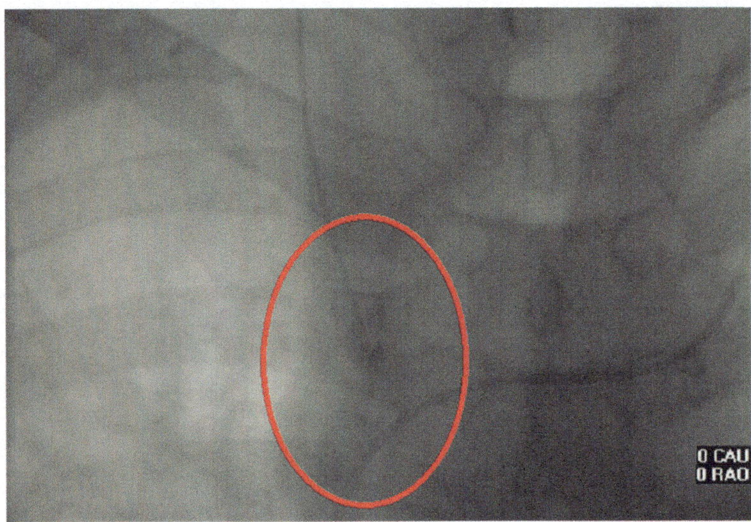

Fig. 28.1 Microcatheter central venogram revealing a severe central stenosis cephalad to a chronically indwelling left chemo port

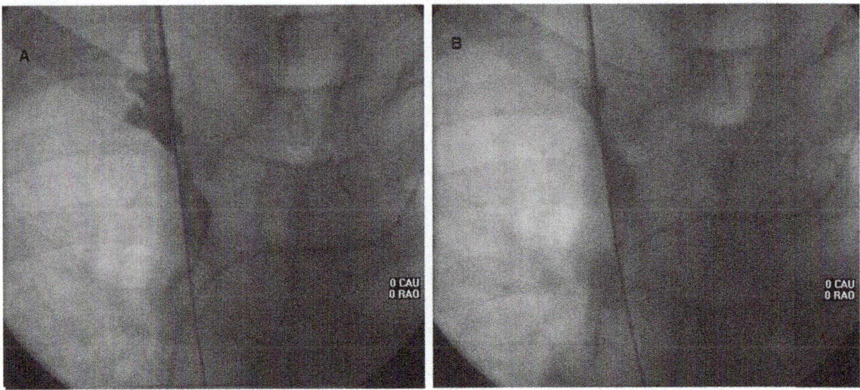

Fig. 28.2 (**a**) Central venogram revealing stenosis (90%) and associated thrombus. (**b**) Post-intervention central venogram revealing improved vasculature for catheter placement

peeled away. Syringe suction revealed good blood flow. Fluoroscopy revealed good placement of the catheter at the right atrial-central vein stenosis (SVC) junction. A 2-0 silk was used to close the insertion point below the clavicle and the base of the neck.

Discussion

The precise mechanism of central vein catheter (CVC)-associated CVS remains largely undefined [1]. A number of factors associated with the catheters themselves, the position of the foreign body against the vessel walls and the uremic milieu and consequent inflammation, all interact to produce CVS. One issue that has been identified is that of turbulent flow and its effect on vascular biology. Turbulence has been shown to cause platelet aggregation and deposition and endothelial hyperplasia [2–4].

Several studies have described a surprisingly high proportion of CVS among HD patients who did not have a history of CVC placement [5, 6]. One study evaluated the use of venography to identify central vein anatomic abnormalities in 69 consecutive patients undergoing percutaneous placement of tunneled RIJ vein catheters. In 29 cases (42%), venography showed evidence of unexpected stenosis and/or angulation of the central veins of sufficient severity to warrant additional fluoroscopy during insertion of the dilators or abandonment of the procedure. Patients who had previously had tunneled internal jugular catheters had more than double the incidence of such abnormalities than those who had not.

Given the above observations and ubiquitous use of catheters at HD initiation in the USA (80% of incident US patients are dialyzed via catheter), the modern interventionalist must be able and willing to salvage as many diseased vessels as possible rather than abandoning the stenosed vein.

References

1. Yevzlin AS. Hemodialysis catheter-associated central venous stenosis. Semin Dial. 2008;21(6):522–7.
2. Glanz S, Gordon DH, Lipkowitz GS, Butt KM, Hong J, Sclafani SJ. Axillary and subclavian vein stenosis: percutaneous angioplasty. Radiology. 1988;168:371–3.
3. Fillinger MF, Reinitz ER, Schwartz RA, Resetarits DE, Paskanik AM, Bruch D, Bredenberg CE. Graft geometry and venous intimal-medial hyperplasia in arteriovenous loop grafts. J Vasc Surg. 1990;11:556–66.
4. Middleton WD, Erickson S, Melson GL. Perivascular color artifact: pathologic significance and appearance on color Doppler US images. Radiology. 1989;171:647–52.
5. Oguzkurt L, Tercan F, Yildirim S, Torun D. Central venous stenosis in haemodialysis patients without a previous history of catheter placement. Eur J Radiol. 2005;55:237–42.
6. Taal MW, Chesterton LJ, McIntyre CW. Venography at insertion of tunnelled internal jugular vein dialysis catheters reveals significant occult stenosis. Nephrol Dial Transplant. 2004;19:1542–5.

Chapter 29
Catheter Placement in a Collateral Vein

Alexander S. Yevzlin

Case Presentation

A 32-year-old female, type I diabetic since early childhood, presented with an infected groin hemodialysis (HD) catheter. She had multiple failed HD and peritoneal dialysis (PD) accesses and was not a transplant candidate due to a high panel reactive antibody (PRA). An attempt was made to evaluate her central vasculature for placement of a catheter through a collateral vein into the right atrium.

After informed consent was obtained, the patient was prepped and draped in the usual sterile fashion overlying the right side of the neck and chest. Using US guidance the left chest collateral was accessed with a 4-French micro-catheter, and angiogram was performed (Fig. 29.1a). This revealed no direct path into the right atrium and multiple collateral vessels on the chest wall. The catheter was redirected to an alternate collateral, and an angiogram revealed a "bird's beak" lesion (Fig. 29.1b).

A hydrophilic guide catheter and a straight hydrophilic wire were inserted into the lesion and aggressively manipulated into the central vasculature. Angioplasty balloon 8×40 mm was advanced over the wire and inflated to 10 atm. Post intervention angiogram is showed in Fig. 29.2. A 12×40 mm angioplasty balloon was then advanced into the SVC and inflated ×3 to 15 atm. Angiogram revealed patent flow from the collateral vessel to the right atrium.

A tunnel was then created from the chest wall below the clavicle to the insertion point of the introducer into the collateral vessel. A 23-cm tunneled HD catheter was inserted into the tunnel. A peel-away sheath was inserted into the right atrium under fluoroscopic guidance. The dilator was removed, and the catheter was inserted into the sheath. The sheath was peeled away. Syringe suction revealed good flow.

A.S. Yevzlin, MD, FASN (✉)
University of Michigan, Ann Arbor, MI, USA
e-mail: asy@medicine.wisc.edu

© Springer International Publishing AG 2017
A.S. Yevzlin et al. (eds.), *Dialysis Access Cases*,
DOI 10.1007/978-3-319-57500-1_29

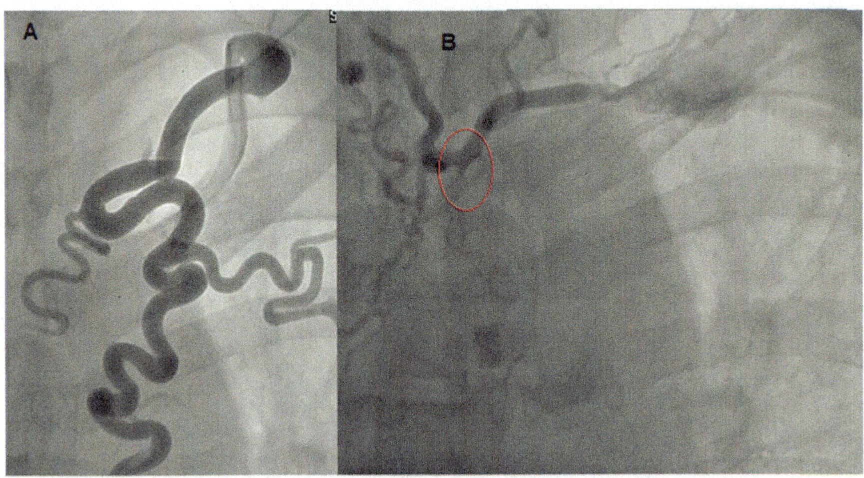

Fig. 29.1 (**a**) Micro-catheter central venogram revealing minimal flow into the central vasculature. (**b**) Redirected catheter injection into alternate collateral reveals a "bird's beak" lesion

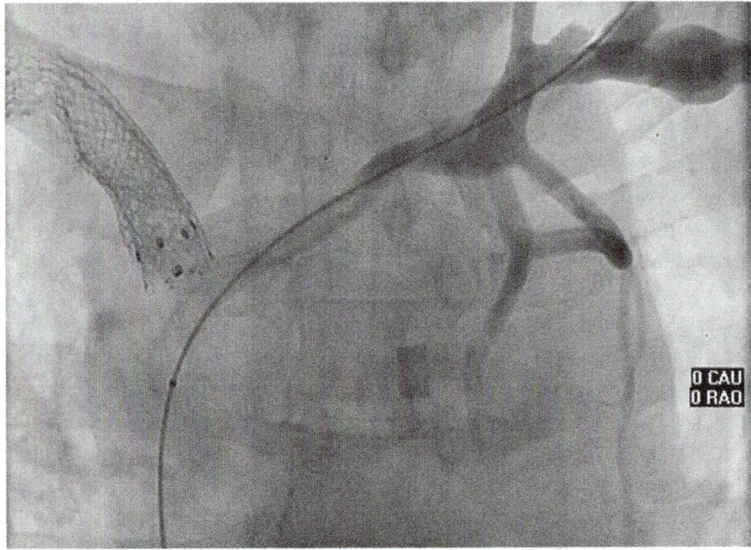

Fig. 29.2 Post intervention angiogram

Fluoroscopy revealed good placement in the right atrial superior vena cava (SVC) border junction. A 2-0 silk was used to close the insertion point and exit sites.

Discussion

The pathophysiology noted in the case described resulted from a combination of related factors. The position of the foreign body against the vessel walls and the uremic milieu and consequent inflammation all interact to produce central vein stenosis (CVS). The mediator between these two factors is turbulent flow and its effect on vascular biology.

Given the importance of central vein catheter (CVC) contact with vessel wall in the pathogenesis of CVS, anatomic considerations that minimize this contact are crucial in selecting the placement location. Perhaps because they are widely regarded as having the least contact with the vessel walls, the right internal jugular (RIJ) CVC is considered the best choice [1–5]. A catheter inserted through the left internal jugular vein (LIJ), for instance, must cross a complex anatomic pathway, which includes the angulations between the LIJ, left brachiocephalic vein, and superior vena cava. These angulations are readily apparent on fluoroscopy and have been implicated in the higher complication rates of left-sided catheters [6–8].

Central venous stenosis is a frequent complication of CVC placement and prolonged use with a complex pathophysiology that remains largely undefined. Despite the impressive impact of national guidelines to improve fistula prevalence, CVC placement rates remain substantially higher than that stipulated by the same guidelines. As the treatment outcomes of CVC-associated CVS, either with percutaneous transluminal angioplasty (PTA) alone, with primary stent placement (PTS), or with surgery, have been disappointing, catheter avoidance remains the best strategy.

References

1. Yevzlin AS. Hemodialysis catheter-associated central venous stenosis. Semin Dial. 2008;21(6):522–7.
2. Beasley C, Rowland J, Spergel L. Fistula first: an update for renal providers. Nephrol News Issues. 2004;18:88–90.
3. DeMeester J, Vanholder R, Ringole S. Factors affecting catheter and technique survival in permanent silicone single lumen dialysis catheters. J Am Soc Nephrol. 1992;3:361A, (Abstract).
4. Schillinger F, Schillinger D, Montagnac R, Milcent T. Post catheterization vein stenosis in haemodialysis: comparative angiographic study of 50 subclavian and 50 internal jugular accesses. Nephrol Dial Transplant. 1991;6:722–4.
5. Cimochowski GE, Worley E, Rutherford WE, Sartain J, Blondin J, Harter H. Superiority of the internal jugular over the subclavian access for temporary hemodialysis. Nephron. 1990;54:154–61.

6. Salgado OJ, Urdaneta B, Colmenares B, Garcia R, Flores C. Right versus left internal jugular vein catheterization for hemodialysis: complications and impact on ipsilateral access creation. Artif Organs. 2004;28:728–33.
7. Wivell W, Bettmann MA, Baxter B, Langdon DR, Remilliard B, Chobanian M. Outcomes and performance of the Tesio twin catheter system placed for hemodialysis access. Radiology. 2001;221:697–703.
8. Lin BS, Kong CW, Tarng DC, Huang TP, Tang GJ. Anatomical variation of the internal jugular vein and its impact on temporary haemodialysis vascular access: an ultrasonographic survey in uraemic patients. Nephrol Dial Transplant. 1998;13:134–8.

Chapter 30
Catheter Placement in the External Jugular Vein

Muhammad Karim and Alexander S. Yevzlin

Case Presentation

A 63-year-old right-handed man with a glomerular filtration rate (GFR) of 7 ml/min was referred to Interventional Nephrology for placement of a tunneled hemodialysis catheter due to the onset of uremic symptoms. Physical examination revealed a prominent REJ. Ultrasound of the neck revealed absence of the right internal jugular (RIJ) and a normal left internal jugular (LIJ). Ultrasound was further used to characterize the veins of the left upper extremity for future access planning, revealing adequate veins for a future left arm arteriovenous fistula (AVF).

After obtaining informed consent, the patient was prepped and draped in the usual sterile fashion overlying the right-hand side of the neck and chest. Ultrasound was used to visualize the veins of the neck, revealing very small right internal jugular (IJ) and very generous right external jugular (REJ) and very small LIJ. Valsalva maneuver demonstrated an increment in the size of the REJ. After local anesthesia, this vein was cannulated under ultrasound guidance using a micropuncture needle. A guidewire was inserted into the lumen of the external jugular (EJ) vein and manipulated under fluoroscopic guidance into the central vasculature. A dilator-introducer combination and 4-French micropuncture were introduced over the wire to the EJ vein. The dilator and wire were removed. Contrast was injected to define venous anatomy, which revealed tortuous EJ vein that communicated with the central vasculature (Fig. 30.1). A guidewire (0.035 cm) was then inserted and navigated

M. Karim, MBBS
Department of Medicine, University of Wisconsin-Madison, Madison, WI, USA

A.S. Yevzlin, MD, FASN (✉)
University of Michigan, Ann Arbor, MI, USA
e-mail: asy@medicine.wisc.edu

© Springer International Publishing AG 2017
A.S. Yevzlin et al. (eds.), *Dialysis Access Cases*,
DOI 10.1007/978-3-319-57500-1_30

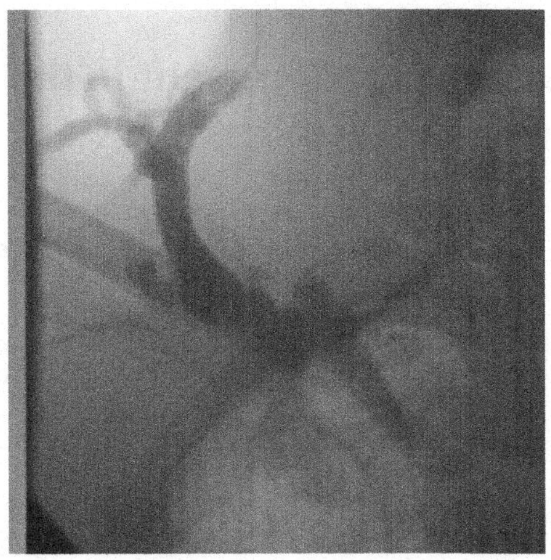

Fig. 30.1 Central venograms via right external jugular vein. Note tortuous course of right external jugular vein

under fluoroscopic guidance into the central vasculature. A 1-cm incision was made at the insertion point of the wire and introducer as well as below the clavicle after 15 ml of subcutaneous lidocaine was administered in the subcutaneous tissues of the neck and chest wall. Blunt dissection was used to create a tunnel from the two incision points, and a catheter was inserted into the tunnel. The 4-French introducer was removed, and serial dilation was used to dilate the track from the skin, through the EJ vein, into the central vasculature under fluoroscopic guidance. A 16.5-Fr dilator-introducer combination was advanced into the central vasculature over the wire under fluoroscopic guidance. The dilator and wire were removed. The catheter was inserted and the sheath peeled away. The tip of the catheter was positioned in the atrium. There were no procedure-related complications.

Discussion

Most commonly, the IJ vein is used as an access site, followed by the femoral vein. The subclavian vein was used historically as a site for dialysis catheter placement, but is now avoided given higher rates of central vein stenosis and obstruction (42%) compared to IJ vein placement [1, 2]. The EJ vein is another site that can be used to obtain vascular access [3]. The vein can also be used as a site for both non-tunneled and tunneled dialysis catheters, with outcomes comparable to both right and left IJ catheter insertions [4–6].

It is important to understand the anatomy of the EJ vein and its relation with the surface anatomy. The EJ vein is formed by the union of the posterior auricular vein and retromandibular vein immediately posterior to the angle of the mandible. The EJ vein runs anterior to the sternocleidomastoid muscle and, at the base of the neck, drains into the subclavian vein. The position of the EJ in relation to the surface anatomy can be estimated by a line drawn from the angle of the mandible to the middle of the clavicle.

In a study of 301 patients with EJ catheter placement, major complications included malposition (2.7%) and hematoma formation (2.7%). There were no severe complications such as nerve injury, pneumothorax, or arterial bleeding [7].

Given the importance of preserving the central vasculature on the ipsilateral side of an AVF or potential AVF, the EJ may be preferable to the IJ in certain circumstances.

References

1. Schillinger D, Montagnac R, Milcent T. Post catheterization vein stenosis in haemodialysis: comparative angiographic study of 50 subclavian and 50 internal jugular accesses. Nephrol Dial Transplant. 1991;6(10):722–4.
2. Barrett N, Spencer S, McIvor J, Brown EA. Subclavian stenosis: a major complication of subclavian dialysis catheters. Nephrol Dial Transplant. 1988;3(4):423–5.
3. Moayedi S, Yang Z, Mack CB. Advanced intravenous access: technique choices, pain scores, and failure rates in a local registry. Am J Emerg Med. 2016;34(3):553–7.
4. Chan M, Gimelli G. How I do it: preferential use of the right external jugular vein for tunneled catheter placement. Semin Dial. 2008;21(2):183–5.
5. Yevzlin AS, Chan MR, Wolff MR. Percutaneous, non-surgical placement of tunneled, cuffed, external jugular hemodialysis catheters: a case report. J Vasc Access. 2007;8(2):126–8.
6. Vats HS, Bellingham J, Pinchot JW, Young HN, Chan MR, Yevzlin AS. A comparison between blood flow outcomes of tunneled external jugular and internal jugular hemodialysis catheters. J Vasc Access. 2012;13(1):51–4.
7. Ishizuka M, Nagata H, Takagi K, Kubota K. External jugular venous catheterization with a Groshong catheter for central venous access. J Surg Oncol. 2008;98(1):67–9.

Chapter 31
Tunneled Femoral Hemodialysis Catheter Placement

Alexander S. Yevzlin

Case Presentation

A 37-year-old female with end-stage renal disease (ESRD) presents with a femoral tunneled hemodialysis catheter that "fell out." The patient has had multiple previous access surgeries and multiple central catheters with known left and right subclavian vein occlusion. She had been dialyzing via right femoral tunneled catheter for the past 3 months with no infectious complications noted. The patient is known to have a contrast allergy.

Ultrasound was used to evaluate the right femoral vein for patency. This revealed a partially thrombosed vessel. Given this ultrasound finding, 50 mg of Benadryl and 100 mg of Solu-Medrol were given prior to start of the procedure, in anticipation of needed contrast angiography.

After informed consent was obtained, the patient was prepped and draped in the usual sterile fashion overlying the right side of the neck and chest. A 15 mL lidocaine was administered in the area overlying the right femoral region. A 22 g needle and 5Fr introducer combination were utilized under US guidance to cannulate the right femoral vein. A wire was inserted under fluoroscopic guidance into the central vasculature through the introducer. Resistance was noted in passing the wire; therefore a venogram was performed (Fig. 31.1). This revealed a patent femoral vein above the inguinal ligament and patent inferior venous cava (IVC).

A 0.035 in J-guidewire was then inserted into the introducer, and a dilator was advanced over the wire into the central vasculature under fluoroscopic guidance. A 1 cm incision was performed at the insertion point of the dilator as well as below the clavicle, and a tunnel was created using blunt dissection from the two incision points.

A.S. Yevzlin, MD, FASN (✉)
University of Michigan, Ann Arbor, MI, USA
e-mail: asy@medicine.wisc.edu

© Springer International Publishing AG 2017
A.S. Yevzlin et al. (eds.), *Dialysis Access Cases*,
DOI 10.1007/978-3-319-57500-1_31

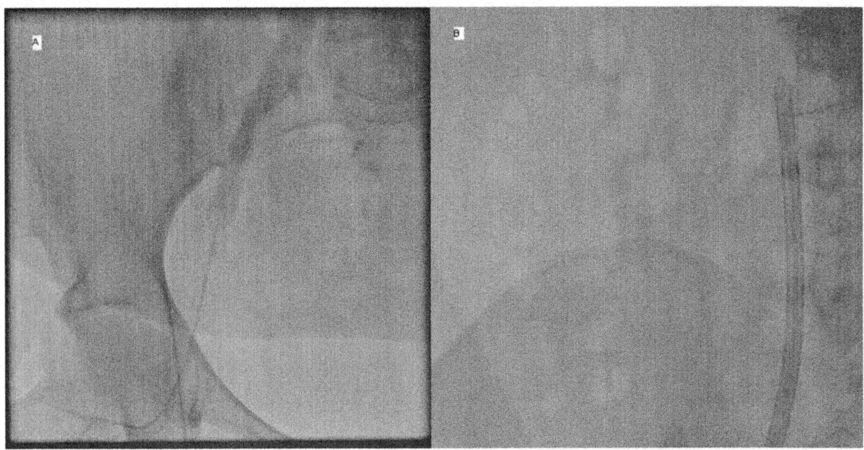

Fig. 31.1 (**a**) Angiogram of femoral vein demonstrated patency above the inguinal ligament. (**b**) Tunneled catheter was inserted into the inferior venous cava

A catheter, a 28 cm Tal Palindrome, was inserted into the lumen of the tunnel. Serial dilation was used to widen the track from the insertion point to the central vasculature. Under fluoroscopic guidance, a dilator-introducer combination was then inserted into the central vasculature. The dilator and wire were removed, and the catheter was inserted into the introducer. The introducer was peeled away. Syringe suction revealed good flow. Fluoroscopy revealed good placement in the IVC. A 2-0 silk was used to close the insertion point.

Discussion

While advanced techniques can often be used to salvage "occluded" access sites, tunneled femoral catheters are the only tenable form of hemodialysis access in a small number of patients. Often times, tunneled femoral vein dialysis catheters are used as a last resort when all other options for a permanent vascular access or thoracic central vein catheter have been exhausted. A study by Maya et al. looked at the outcomes of tunneled femoral catheters [1]. The authors used a prospective, computerized vascular access database to identify all tunneled femoral dialysis catheters placed at the University of Alabama at Birmingham during a 5-year period.

Twenty-seven subjects received a tunneled femoral dialysis catheter, accounting for 1.9% of all tunneled catheters placed, during the study period. The primary catheter patency (time from placement to exchange) was substantially shorter for femoral catheters than for internal jugular dialysis catheters (median survival, 59 vs. >300 days, $P < 0.0001$). Interestingly, infection-free survival was similar for both groups

($P = 0.66$). Seven patients with femoral catheters (or 26%) developed an ipsilateral deep vein thrombosis, but catheter use was possible with anticoagulation.

In contrast to commonly held belief about tunneled femoral catheters, this study demonstrated no difference in infection rates compared to thoracic tunneled catheters. Tunneled femoral dialysis catheters were found to have a substantially shorter primary patency, however. This may be explained by the fact that blood is drawn by gravity into the lumen of the catheter from above in the case of femoral lines.

Reference

1. Maya ID, Allon M. Outcomes of tunneled femoral hemodialysis catheters: comparison with internal jugular vein catheters. Kidney Int. 2005;68(6):2886–9.

Chapter 32
Transhepatic Catheter Placement

Jason W. Pinchot and Christopher M. Luty

Abbreviations

IVC	Inferior vena cava
NKF KDOQI	The National Kidney Foundation Kidney Disease Outcomes Quality Initiative
TPN	Total parenteral nutrition

Case Presentation

The patient was a 22-year-old female with history of chronic intestinal pseudo-obstruction who underwent multivisceral (liver and small intestine) organ transplant 10 years ago. Before and subsequent to her transplant, she had innumerable central venous catheters and hemodialysis catheters placed for total parenteral nutrition (TPN) administration and dialysis, respectively. A number of these lines were removed soon after placement secondary to frequent central line-associated bloodstream infections and sepsis. Past evaluations had shown that both innominate veins were occluded (Fig. 32.1). Moreover, the patient had known chronic total occlusion of the femoral veins and infrahepatic inferior vena cava (Fig. 32.2). She was referred to the interventional radiology section for placement of a tunneled dialysis catheter.

J.W. Pinchot, MD (✉)
Department of Radiology, Vascular an Interventional Radiology Section, University of Wisconsin School of Medicine and Public Health, 600 Highland Avenue, CSC D4/346, Madison, WI 53792, USA
e-mail: jpinchot@uwhealth.org

C.M. Luty, MD
University of Wisconsin-Madison, Department of Radiology, Vascular and Interventional Radiology, Madison, WI, USA

© Springer International Publishing AG 2017
A.S. Yevzlin et al. (eds.), *Dialysis Access Cases*,
DOI 10.1007/978-3-319-57500-1_32

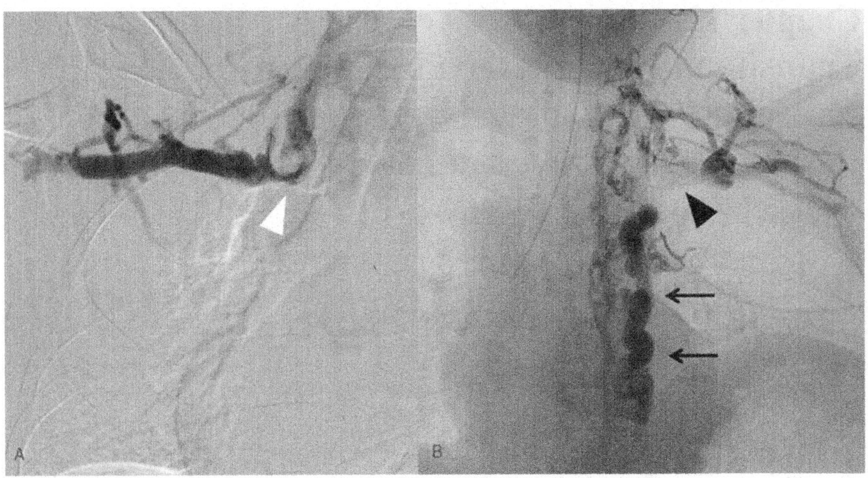

Fig. 32.1 Conventional venography reveals chronic total occlusion of the *right innominate vein* (Image (**a**) *white arrowhead*) and *left subclavian vein* (Image (**b**) *black arrowhead*). Note presence of robust posterior mediastinal collateral veins (Image (**b**) *black arrows*)

Fig. 32.2 Chronic total occlusion of the right common iliac vein and inferior vena cava (IVC). A short segment of suprahepatic IVC is reconstituted from a number of ascending lumbar and paraspinal collateral veins

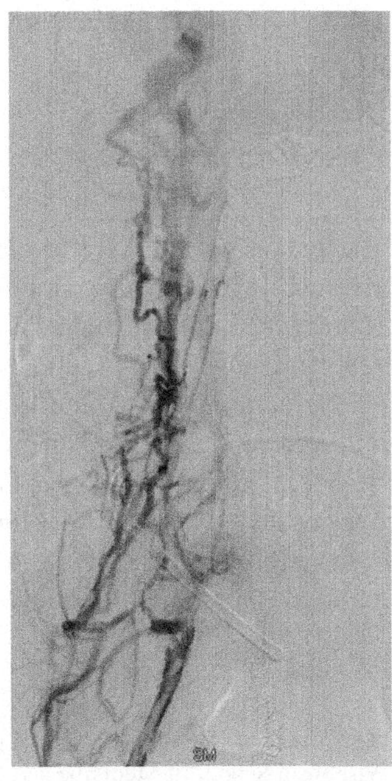

Physical examination revealed multiple healed incisions overlying the deltopectoral groove on both sides, the result of numerous prior tunneled subclavian and internal jugular venous catheters. Superficial collateral veins were present on the anterior abdominal wall, in keeping with known iliocaval venous occlusion. With this in mind, a transhepatic catheter was felt to be the best approach for central venous access.

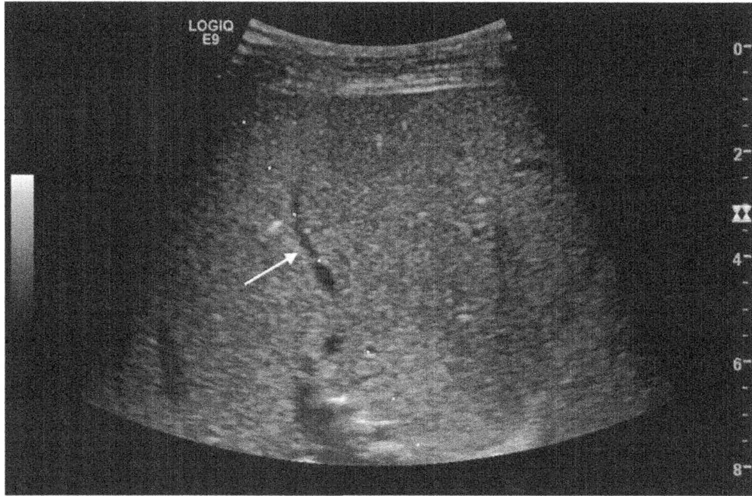

Fig. 32.3 Pre-procedure grayscale ultrasound image demonstrates the planned transhepatic route of needle puncture into the middle hepatic vein

Fig. 32.4 Conventional venogram confirms access into the peripheral aspect of the middle hepatic vein

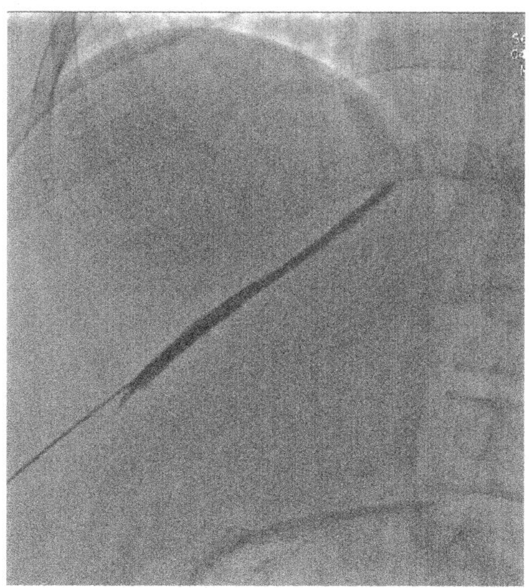

Color Doppler ultrasound interrogation of the liver demonstrated wide patency of the middle and right hepatic veins with normal flow into the suprahepatic inferior vena cava. The right upper quadrant of the patient's abdomen was prepped and draped in standard sterile manner. Buffered 1% lidocaine was used for local anesthesia of the skin, subcutaneous tissues, and liver capsule. The middle hepatic vein was targeted for access (Fig. 32.3). Using a 22-gauge needle and AccuStick introducer sheath, the middle hepatic vein was successfully cannulated, and a middle hepatic venogram (Fig. 32.4) was performed. This confirmed access into the peripheral aspect of the middle hepatic vein. A guidewire was passed through the introducer

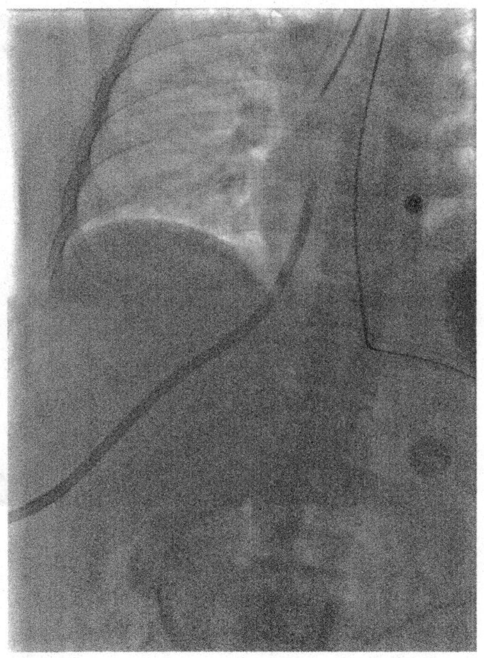

Fig. 32.5 Tunneled 19-cm dialysis catheter in place. Note appropriate position of the catheter tip in the *right atrium* above the level of the inferior cavoatrial junction

sheath into the right atrium. The introducer sheath was exchanged over the guidewire for sequential soft tissue dilators followed by placement of a 16-Fr valved peel-away sheath. Additional local anesthesia was then performed inferior and lateral to the venous entry site, and a subcutaneous tunnel was fashioned. A 14.5-Fr 28-cm cuffed hemodialysis catheter was then pulled through the tunnel and brought out at initial transhepatic venous entry site. The inner dilator and guidewire were removed from the peel-away sheath, and the catheter was passed through the sheath into the right atrium (Fig. 32.5).

At the completion of the procedure, both catheter ports aspirated and flushed without difficulty. The patient was subsequently transferred to the dialysis center where her catheter was used without complication.

Discussion

A tunneled cuffed hemodialysis catheter should be placed in a vein that is easily accessible using sonographic and fluoroscopic guidance. The internal jugular veins are generally favored as the initial choice for central venous access [1]. If both of the internal jugular veins become occluded, alternative access sites must be explored. Choices include the external jugular and femoral veins. The subclavian veins can be considered but are generally avoided, in accordance with the National Kidney Foundation Kidney Disease Outcomes Quality Initiative (NKF KDOQI) guidelines, since catheterization results in a high risk of central stenosis which would generally preclude the use of the entire ipsilateral arm for vascular access. As with this case, prolonged central venous catheter use manifests a high rate of venous thrombosis and chronic total occlusion with exhaustion of even these less routine access sites. Unconventional approaches to central venous access—translumbar and transhepatic catheter placement—may be entertained only when all other endovascular options for central venous access have been exhausted. To be sure, these sites require considerable technical expertise for catheter placement, and maintenance of catheters at these sites may be somewhat more problematic [2]. Transhepatic catheter placement has become the favored approach to central venous access in patients with coexistent chronic occlusion of the infrahepatic inferior vena cava.

Percutaneous transhepatic puncture of a hepatic vein for dialysis access was first described by Po et al. in a case report in 1994 [3]. Since this time, subsequent retrospective studies have detailed long-term safety, efficacy, and complications inherent to transhepatic central venous access [4–6]. While they are well tolerated, transhepatic catheter malfunctions are almost all due to the natural excursion of the liver during the respiratory cycle resulting in catheter misplacement, kinking, and poor blood flow [5]. Most investigators agree that placement is associated with low morbidity but note high rates of catheter-related maintenance. In a 2003 series of 12 patients with 36 lines reported by Stavropoulos et al., the mean time of the catheters in situ was mere 24.3 days [4]. More recently, a study from 2013 reviewed their single-center experience in which total access site patency was seen up to 790 days when taking in account two line replacements for thrombosis and displacement [7].

The technique for transhepatic access is rather straightforward. Prior to the procedure, ultrasonography of the right upper quadrant is performed to document patency of the right or middle hepatic vein. Using a standard sterile technique, buffered 1% lidocaine is administered for local anesthesia taking special care to anesthetize the superficial and deep soft tissues including the liver capsule. Under ultrasound guidance, a 21-gauge, 15-cm-long needle is advanced into the targeted vein from an anterior subcostal or mid-axillary intercostal approach. The subcostal approach may help to limit future catheter malposition as it minimizes excursion with respirations [5]. Transhepatic cannulation of a hepatic vein is preferred over direct inferior vena cava puncture because it permits a longer intravascular tract and decreases the chance of migration out of the vessel [2]. A 0.018-inch platinum-tipped mandril guidewire is then advanced through the needle and into the right atrium. Intravascular catheter length is measured, and the proper catheter is selected in standard fashion. Typically, this will be in the 20–30-cm range. The initial access needle is exchanged over the guidewire for a coaxial transitional sheath (AccuStick system, Boston Scientific, Natick, Massachusetts), which permits replacement of the 0.018-inch guidewire with a 0.035-inch guidewire. In obese patients or in those with cirrhosis, a stiff guidewire may be necessary to facilitate transhepatic passage of the peel-away sheath. Additional local anesthesia is administered inferior and lateral to the venous entry site, and a subcutaneous tunnel is fashioned. The hemodialysis catheter is pulled through the tunnel and brought out at the venotomy. Over the guidewire, the transitional dilator is exchanged for an appropriately sized valved peel-away sheath, which is advanced in to the hepatic vein. Pre-dilation with sequential rigid dilators may be necessary. Once the sheath is in place, the inner dilator and guidewire are removed, and the catheter is introduced through the sheath and into the central venous circulation. Some interventionalists opt to keep a stiff hydrophilic guidewire in place and then advance the catheter through the sheath and over the guidewire into the hepatic vein until the tip lies within the right atrium [8]. Both catheter ports are flushed, heparinized, and secured. The initial venous access site is closed using interrupted sutures, Steri-Strips or Dermabond.

Pearls

- Transhepatic catheter placement is a viable means of central access when conventional routes have been exhausted.
- The technique for transhepatic access is straightforward, especially when the operator is skilled with ultrasound.
- While catheter malfunctions are more frequent than those encountered with conventional central venous access, durable access can be achieved with routine maintenance.

References

1. Cimchowski GE, Workey E, Rutherford WE, Sartain J, Blondin J, Harter H. Superiority of the internal jugular over the subclavian access for temporary dialysis. Nephron. 1990;54(2): 154–61.
2. Pinchot JW. Unconventional venous access: percutaneous translumbar and transhepatic venous access for hemodialysis. In: Yevzlin AS, Asif A, Salman L, editors. Interventional nephrology: principles and practice. New York: Springer; 2014. p. 49–55.
3. Po CL, Koolpe HA, Allen S, Alvez LD, Raja RM. Transhepatic PermCath for hemodialysis. Am J Kidney Dis. 1994;24(4):590–1.
4. Stavropoulos SW, Pan JJ, Clark TW, Soulen MC, Schlansky-Goldberg RD, Itkin M, Trerotola SO. Percutaneous transhepatic venous access for hemodialysis. J Vasc Interv Radiol. 2003; 14:1187–90.
5. Smith TP, Ryan JM, Reddan DN. Transhepatic catheter access for hemodialysis. Radiology. 2004;232:246–51.
6. Younes HK, Pettigrew CD, Anaya-Ayala JE, Soltes G, Saad WE, Davies MG, et al. Transhepatic hemodialysis catheters: functional outcome and comparison between early and late failure. J Vasc Interv Radiol. 2011;22:183–91.
7. Herscu G, Woo K, Weaver FA, Rowe VL. Use of unconventional dialysis access in patients with no viable alternative. Ann Vasc Surg. 2013;27(3):332–6.
8. Weeks SM. Unconventional venous access. Tech Vasc Interv Radiol. 2002;5(2):114–20.

Chapter 33
Anterior Jugular Vein Tunneled Dialysis Catheter

Gerald A. Beathard

Case Presentation

The patient was a 67-year-old male who had been on hemodialysis for 8 years. During this period of time, he had had multiple central venous catheters inserted. Past evaluations had shown that both internal jugular veins were occluded. The patient was referred to the dialysis access center for the placement of a tunneled dialysis catheter.

On physical examination, the patient was found to have multiple old nonfunctional arteriovenous access sites on both arms. Multiple scars were present over both subclavicular areas at old exit sites of previous dialysis catheters. Examination of the cervical area with Doppler ultrasound revealed the presence of a large dilated superficial vein slightly to the left of the cervical midline. The size of this vein increased when the patient performed a Valsalva maneuver.

After the patient was prepped and draped, the previously noted vein was cannulated using a 21 gauge Angiocath, and an angiogram was performed (Fig. 33.1). This image showed that this vein was the left anterior jugular vein, which drained into the central veins at the confluence of the left subclavian and brachiocephalic veins.

A guidewire was inserted and manipulated to enter the left brachiocephalic vein (Fig. 33.2). This was followed by a vascular catheter, which was used to perform in a second angiogram to confirm the position (Fig. 33.3). A 32 cm cuffed dialysis catheter was inserted over the guidewire and placed so that its tip was within the right atrium. Because of the angle of entry at the cannulation site, the catheter was tunneled to the right (Fig. 33.4).

G.A. Beathard, MD, PhD (✉)
Lifetime Vascular Access, University of Texas Medical Branch,
5135 Holly Terrace, Houston, TX 77056, USA
e-mail: gbeathard@msn.com

© Springer International Publishing AG 2017
A.S. Yevzlin et al. (eds.), *Dialysis Access Cases*,
DOI 10.1007/978-3-319-57500-1_33

Fig. 33.1 Initial
angiogram. *A*Anterior
jugular vein, *B* left
brachiocephalic vein

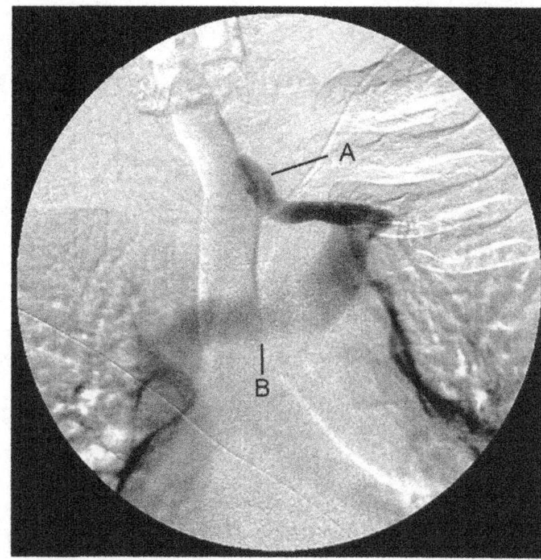

Fig. 33.2 Course of
guidewire from
cannulation site to superior
vena cava. *A* Guidewire

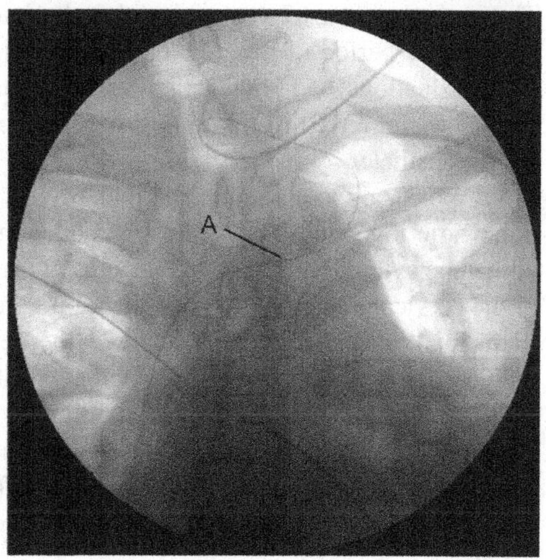

After placement, the catheter functioned well when tested with a 10 mL syringe. The patient was returned to the dialysis facility where the catheter was used without difficulty.

Discussion

The jugular venous system constitutes the primary venous drainage of the head and neck. It includes a subfascial venous system, formed by the right and left internal jugular veins, and a superficial or subcutaneous one, formed by the two anterior, two

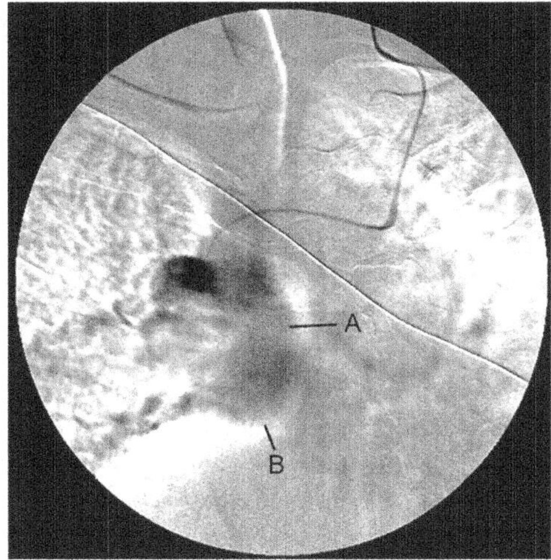

Fig. 33.3 Angiogram performed through vascular catheter. *A* Superior vena cava, *B* right atrium

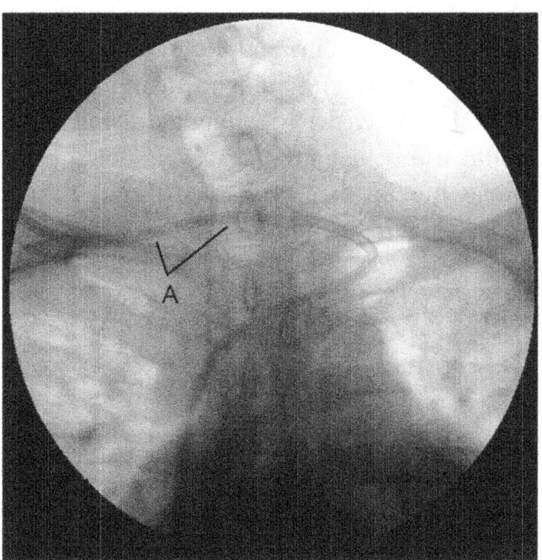

Fig. 33.4 Tunneled catheter in place. *A* Tunneled catheter

external jugular veins, and two posterior external jugular veins. The internal jugular veins provide the major route for cervical drainage and are used extensively for the placement of central venous dialysis catheters. With chronic repetitive catheter use; however, these veins frequently become stenosed and are lost. This causes the development of collateral flow to the other cervical veins, primarily the external jugular veins, the size of which tends to be inversely related to the size of the internal jugular veins.

The external jugular veins are frequently used for the placement of dialysis catheters in cases in which the internal jugular veins are not accessible. Unfortunately, this can also result in the loss of these venous conduits. The anterior jugular veins frequently become enlarged in this situation. In fact, this is the major means by which the anterior jugular veins even become apparent and available as an alternative route for catheter placement.

The anterior jugular vein typically begins near the hyoid bone by the confluence of several superficial veins from the submaxillary region. It descends between the median line and the anterior border of the sternocleidomastoid muscle. At the lower part of the neck, it passes beneath that muscle to open into the termination of the external jugular or, in some instances, into the subclavian vein. Typically, a right and a left vein are present. However, in some cases it is single [1]. These veins drain into the subclavian (54%) or into the external jugular vein (46%) [2]. Just above the sternum, the two anterior jugular veins communicate by a transverse trunk, the jugular venous arch, which also receives tributaries from the inferior thyroid veins.

As vascular access catheter sites become progressively depleted, alternate anatomic locations must be sought. The interventionalist caring for a population of dialysis patients must be familiar with the venous anatomy of the neck and be prepared to utilize any vein that allows for entry into the superior vena cava.

The superficial jugular veins present good candidates for catheter placement in cases in which the internal jugular veins have neen lost. Because of the superficial nature of these veins and their compressibility, they are at times difficult to cannulate with ultrasound guidance using a standard needle. The author has found that it is frequently easier to cannulate using an Angiocath using the technique that one would use in cannulating a vein for a peripheral venipuncture. A micropuncture guidewire can then be passed through this device and proceed with the standard micropuncture technique thereafter.

References

1. Werner JD, Siskin GP, Mandato K, Englander M, Herr A. Review of venous anatomy for venographic interpretation in chronic cerebrospinal venous insufficiency. J Vasc Interv Radiol. 2011;22(12):1681–90; quiz 1691.
2. Deslaugiers B, Vaysse P, Combes JM, Guitard J, Moscovici J, Visentin M, et al. Contribution to the study of the tributaries and the termination of the external jugular vein. Surg Radiol Anat. 1994;16(2):173–7.

Chapter 34
Transatrial Internal Jugular Dialysis Catheter in Inferior Vena Cava

Gerald A. Beathard

Case Presentation

The patient was a 49-year-old catheter-dependent hemodialysis patient. He was referred to the vascular access facility because of catheter dysfunction which had been recurrent. At the time the old right internal jugular tunneled catheter was removed, radiocontrast was injected to determine if a fibrin sheath was present. This was found not to be the case. A new 28 cm tunneled catheter was inserted without difficulty. When blood flow was tested using a 10 mL syringe, there was a significant problem. The catheter was repositioned several times in an attempt to obtain good blood flow; however, there was no improvement.

A decision was made to insert a longer catheter and pass it down into the inferior vena cava (Fig. 34.1) (IVC). A 48 cm step-tip tunneled catheter was inserted over a guidewire. Fluoroscopic evaluation of the catheter tip revealed that it was well below the level of the diaphragm. When tested using a 10 mL syringe, blood flow in the catheter was found to be good. The patient was returned to the dialysis facility and dialyzed without difficulty. The catheter was connected in a reversed manner, e.g., the arterial blood line was connected to the venous side of the catheter and the venous bloodline to the arterial.

G.A. Beathard, MD, PhD (✉)
Lifetime Vascular Access, University of Texas Medical Branch,
5135 Holly Terrace, Houston, TX 77056, USA
e-mail: gbeathard@msn.com

© Springer International Publishing AG 2017
A.S. Yevzlin et al. (eds.), *Dialysis Access Cases*,
DOI 10.1007/978-3-319-57500-1_34

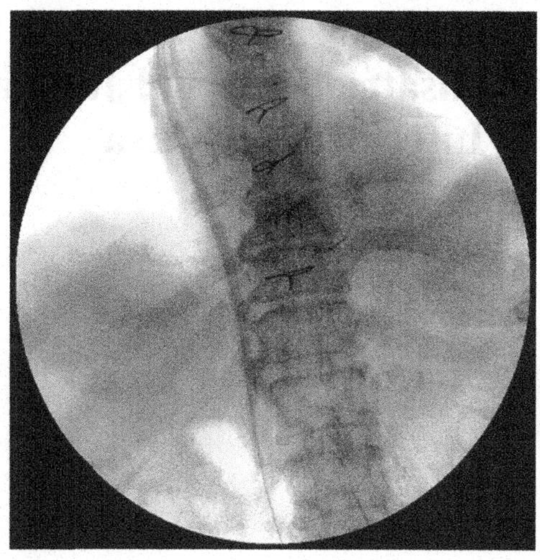

Fig. 34.1 Appearance of dialysis catheter with tip in inferior vena cava

Discussion

Once a tunneled dialysis catheter is in place, it should be tested with a 10 cc syringe repetitively (4 or 5 times) to judge the adequacy of blood flow. The ability to fill the syringe within 2 s is equivalent to 300 mL/min blood flow. Any drag or "stutter" when withdrawing the blood is an indication that the catheter tip position needs to be adjusted, once any possible mechanical issues have been eliminated. There are times, however, when even with multiple adjustments, optimal blood flow cannot be obtained and all of the obvious reasons for the problem have been ruled out. In the author's experience, this is more likely to occur in a situation in which the patient has had a dialysis catheter for a prolonged period of time. The etiology for this problem frequently cannot be defined (once the obvious have been ruled out); however, the suspicion is that it is secondary to the presence of fibrin strands within the atrium as a result of previous catheters that cause intermittent obstruction of the catheter tip. In this situation, the best alternative is to advance the tip of the catheter into the inferior vena cava [1, 2].

It should be noted that the primary determinant of catheter blood flow resistance is the inner diameter of the catheter. It is inversely related to the radius raised to the fourth power (r^4). While length also plays a role, resistance is only directly proportionate to the length. This being the case, increasing the length does not add very much to the negative pressure registered by the dialysis machine when the catheter is being used. Actually a 19% increase in the diameter of the catheter will offset a doubling of its length as far as resistance is concerned.

Very little has been published related to the cervical placement of a tunneled dialysis catheter with the tip located within the inferior vena cava. Only anecdotal reports are available [1, 2], but the experience that has been reported has been favorable. The only significant risk to this type of catheter placement that has been recognized is that which may occur secondary to catheter retraction. When the patient moves from a supine to a sitting position, the catheter generally retracts. This change often amounts to 3–4 cm and sometimes more. If the catheter that has been extended into the inferior vena cava from above is too short, the tip can retract back into the atrium when the patient assumes an erect position and cross the tricuspid valve when the patient's position again changes to supine. To minimize this risk, a catheter which passes through the atrium should extend at least 8–10 cm into the inferior vena cava. Because of the direction of blood flow in the inferior vena cava relative to the catheter should be connected backward when used in order to reduce the possibility of recirculation.

References

1. Powell S, Belfield J. Complex central venous catheter insertion for hemodialysis. J Vasc Access. 2014;15(Suppl 7):S136–9.
2. Renaud C, Seow Y, Teh H. Transjugular tunnelled dialysis catheter tip placement into the inferior vena cava upper segment after length overestimation. J Vasc Access. 2014;16:72–5 [Epub ahead of print].

Chapter 35
Internal Jugular Catheter Insertion with No Vein Apparent on Ultrasound

Gerald A. Beathard

Case Presentation

The patient was a 48-year-old male with a 4-year history of hemodialysis. The patient had had numerous central venous catheters during this period of time. He is referred to the vascular access center for the placement of a tunnel dialysis catheter.

Examination of the patient revealed several old abandoned peripheral arteriovenous accesses, none of which were functional. A number of scars were present on the anterior chest; it sites a previous central venous catheter placement. Ultrasound examination of the potential cannulation sites for the right and left internal jugular veins revealed that they were not apparent (Fig. 35.1a). Examining at a higher level on the right side of the neck, a patent internal jugular vein was identified (Fig. 35.1b).

The vein identified at the higher level in the patient's neck was cannulated and an angiogram was performed. This showed what appeared to be total occlusion of the lower internal jugular vein. However, it was possible to pass a 0.35 inch guidewire through this site down through the heart and into the inferior vena cava (Fig. 35.2). And 8 × 4 angioplasty balloon was inserted through a 6 F sheath and the stenotic area was completely dilated (Fig. 35.3). An angiogram performed at this time showed radiocontrast flowing into the superior vena cava and right atrium (Fig. 35.4).

The angioplasty balloon was positioned at the optimal site for cannulation of the internal jugular vein and inflated to a low pressure. Using fluoroscopy guidance, the most cephalad aspect of the inflated balloon was cannulated with a micropuncture needle. The 0.018 inch microguidewire was inserted into the balloon under direct fluoroscopic guidance and allowed to coil so as to become entrapped within the balloon (Fig. 35.5). The balloon was then advanced over the guiding guidewire to carry

G.A. Beathard, MD, PhD (✉)
Lifetime Vascular Access, University of Texas Medical Branch,
5135 Holly Terrace, Houston, TX 77056, USA
e-mail: gbeathard@msn.com

© Springer International Publishing AG 2017
A.S. Yevzlin et al. (eds.), *Dialysis Access Cases*,
DOI 10.1007/978-3-319-57500-1_35

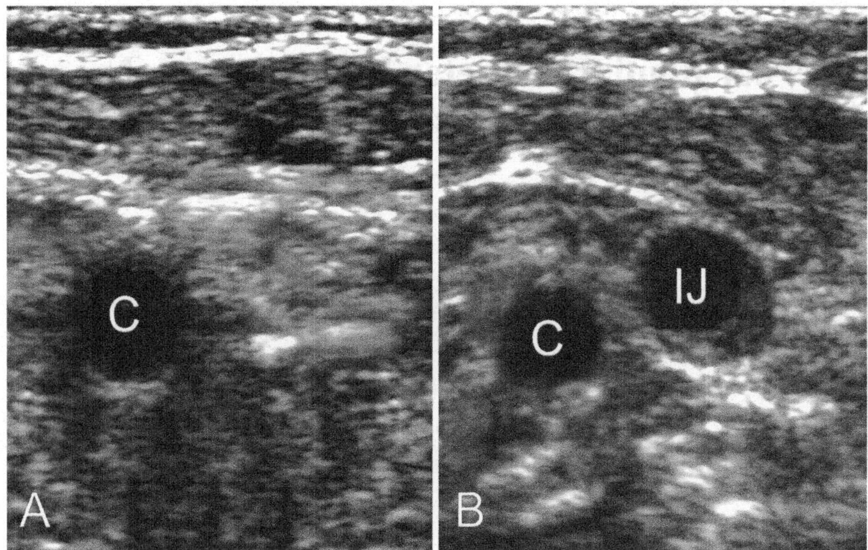

Fig. 35.1 Ultrasound appearance of right internal jugular vein area. (**a**) Appearance of supraclavicular area. Note the absence of the internal jugular vein. *C* Carotid artery. (**b**) Appearance at a higher level. Note the presence of the internal jugular vein. (c - carotid artery, IJ - internal jugular vein)

Fig. 35.2 Initial angiogram with guidewire. *A* Patent internal jugular vein, *B* site of stenosis, *C* guidewire in superior vena cava

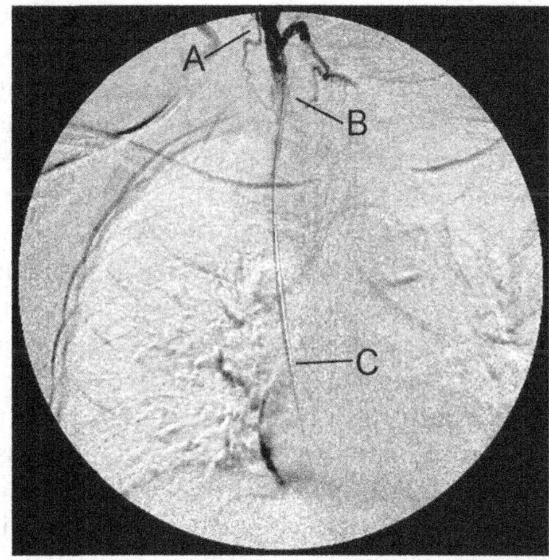

the microguidewire down into the superior vena cava (Fig. 35.6). When it reached this point, the microguidewire was secured manually while the angioplasty balloon was advanced further so as to extract the guidewire from the balloon. Once this had been accomplished, the angioplasty balloon and 0.035 guidewire were extracted. The sheath was left in place until catheter placement had been completed.

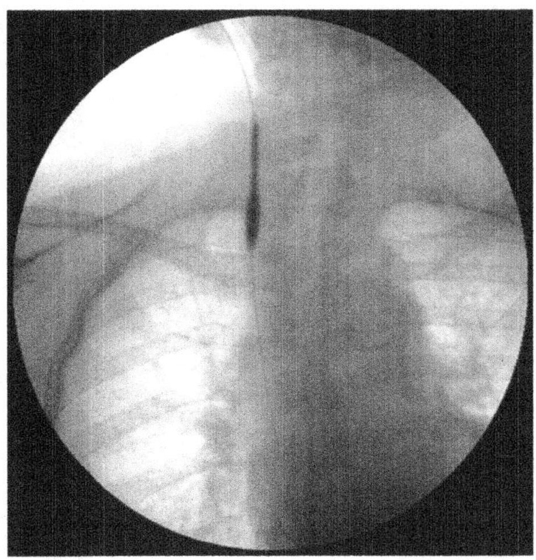

Fig. 35.3 Angioplasty of stenotic zone

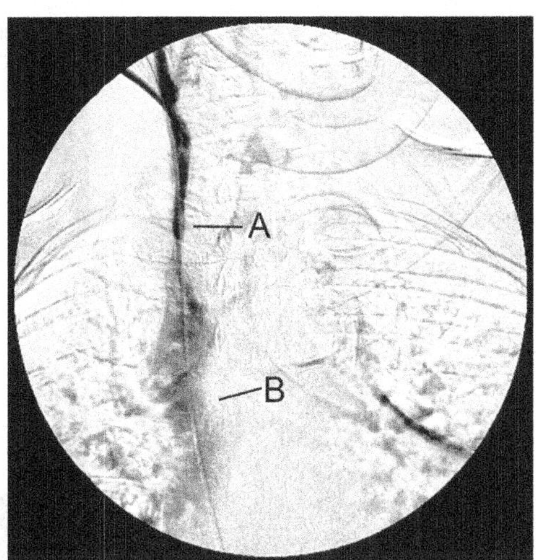

Fig. 35.4 Post-angioplasty angiogram with the guidewire in place

With the microguidewire optimally placed, the standard procedure for tunneled catheter insertion was followed. The catheter was successfully placed, was flushed well, has been tested with a 10 mL syringe, and was used successfully for dialysis.

Discussion

Unfortunately, in many dialysis patients the internal jugular vein is lost due to repetitive catheter use. However, when the jugular vein is not apparent upon ultrasound examination, it does not always mean that it is not there. It only means that it is not

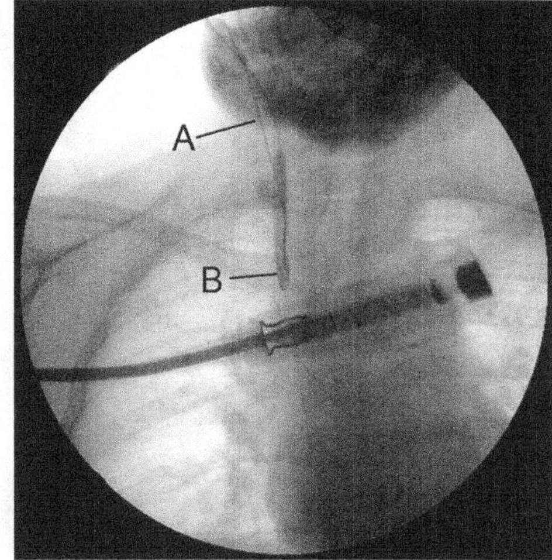

Fig. 35.5 Cannulation of angioplasty balloon and insertion of micro guidewire into the balloon. *A* Cannulation needle, *B* guidewire within the angioplasty balloon

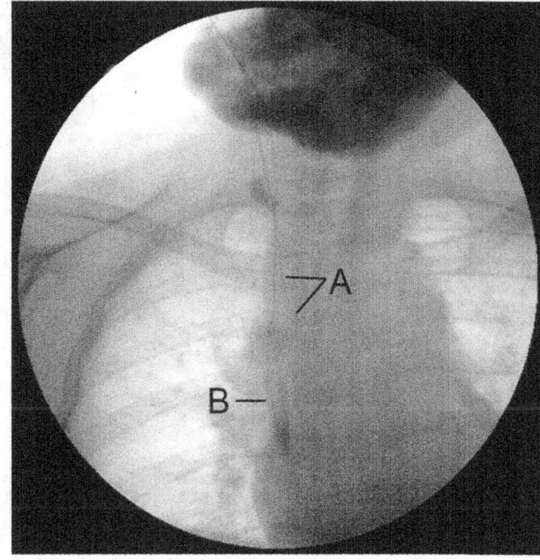

Fig. 35.6 Advancing an angioplasty balloon to pull microguidewire into superior vena cava. *A* Microguidewire in superior vena cava, *B* microguidewire in balloon

apparent. It may be extremely stenotic and too small to be apparent. When the vein is not visualized at the usual site for cannulation, it is worthwhile to examine the neck at a higher level to see if it is visible there. If it is visible, it should be cannulated at this higher point with a micropuncture needle to determine if it is possible to pass the microguidewire down into the central veins from this level. It may also be possible to gain access to an extremely stenotic internal jugular vein or a collateral in the neck from below using a femoral approach. If the guidewire enters the central veins from above or if it passes above the clavicle from below, it may be possible to achieve cannulation at the optimal site using the technique of balloon-guidewire entrapment.

This technique is best performed under direct fluoroscopic guidance, and due to the degree of control of the needle and microguidewire that is required, it is virtually impossible for the operator not to have a hand in the fluoroscopic field as seen in the images associated with this case.

Modifications to this procedure as described are possible. Once the internal jugular vein is cannulated and a guidewire has been inserted into the vena cava, the remainder of the procedure could very easily be done using ultrasound guidance. Additionally, it might be possible, once the vein has been dilated, to cannulate it at the optimal site directly using ultrasound guidance.

Using a target for cannulation of a vessel is a very valuable technique. All interventionalist working with dialysis vascular access should possess the skill. Other targets have been used to guide cannulation such as a snare or a dilator. In the case of the dilator, the guidewire is not entrapped; it just serves as an identifiable target for the cannulating needle.

Chapter 36
Wiring the Tunnel to Replace a Tunneled Dialysis Catheter

Gerald A. Beathard

Case Presentation

The patient was a 38-year-old cachectic female who had been dialyzing with a cuffed tunneled dialysis catheter for the past 2 months. The patient was referred to the dialysis access facility to replace her catheter which had become dislodged at some point during the previous night. The patient had awoken to find the loose catheter in the bed. Examination of the old catheter exit site revealed that it was clean without evidence of inflammation or infection. The site was prepped and draped. A 5 Fr dilator was gently inserted into the exit site directed along the apparent course of the previous catheter. A guidewire was inserted through the dilator, advanced up the tunnel of the old catheter into the central circulation. A new catheter was then inserted over the guidewire and sutured in place. The catheter flowed well when tested with a 10 mL syringe. The patient was referred to the dialysis facility where she dialyzed without difficulty (Figs. 36.1, 36.2, 36.3, 36.4, 36.5, and 36.6).

Discussion

There are several situations in which a tunneled dialysis catheter may be inadvertently removed in a patient who is continued to be catheter-dependent. This may happen at the dialysis facility due to traction that is placed on the catheter if the blood lines are not positioned and taped in a manner that avoids such a problem. While this is more likely to occur in a newly placed catheter, it can occur later if for some reason the retention cuff is not well attached. It can also occur during a

G.A. Beathard, MD, PhD (✉)
Lifetime Vascular Access, University of Texas Medical Branch,
5135 Holly Terrace, Houston, TX 77056, USA
e-mail: gbeathard@msn.com

© Springer International Publishing AG 2017
A.S. Yevzlin et al. (eds.), *Dialysis Access Cases*,
DOI 10.1007/978-3-319-57500-1_36

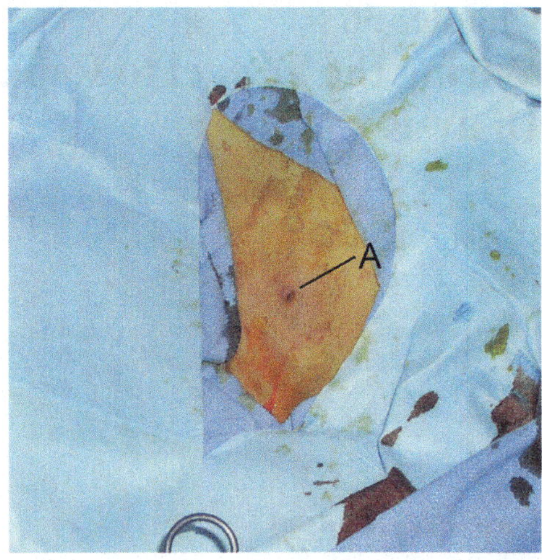

Fig. 36.1 Appearance of old catheter exit site after it has been prepped and draped for new catheter placement. *A* Old catheter exit site

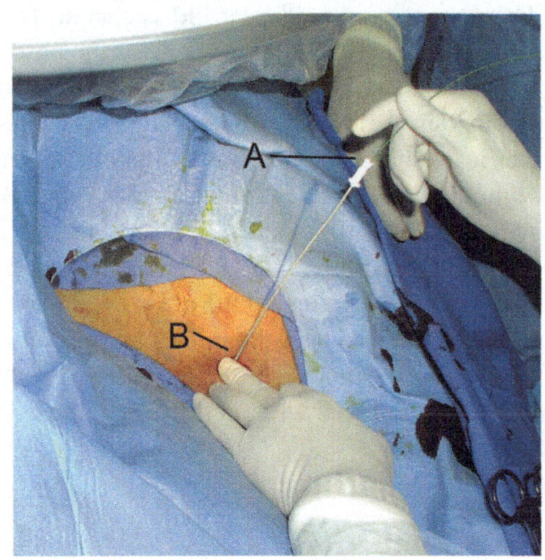

Fig. 36.2 Dilator inserted into the old exit site, beginning to advance the guidewire. *A* Guidewire being inserted into the dilator, *B* dilator inserted into the old exit site

catheter exchange if the old catheter is inadvertantly extracted prior to inserting a guidewire. Occasionally, a patient presents with their catheter in hand having founded it in their bed when they awoke that morning.

Over a relatively short period of time, the subcutaneous tunnel leading to the venous entry site develops a fibroepithelial sheath-like lining that extends up to the venous entry site. This is actually continuous with the fibrin sheath that frquently develops and extends from the venous entry site downward. If the previous catheter was in place long enough for this tunnel to develop, it is possible to utilize it to insert a new catheter through the existing site. With time, there is a

Fig. 36.3 Guidewire advanced up to the vein entry site. *A* Tip of guidewire

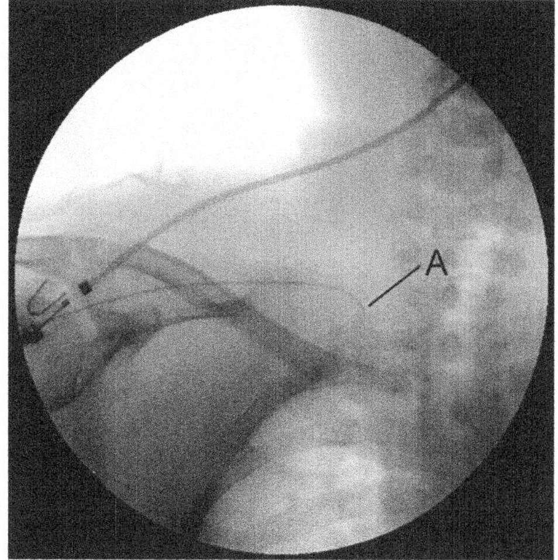

Fig. 36.4 Guidewire advanced into the vein entry site. *A* Tip of guidewire

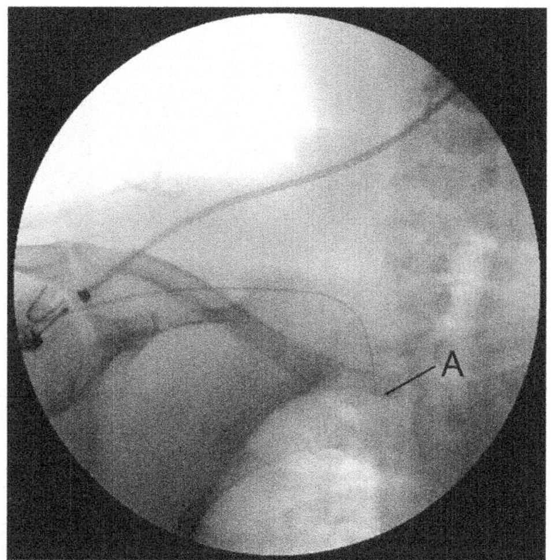

tendency for thrombus to form in the more central portion of the tunnel. This opportunity is lost as this thrombus begins to organize. In general, little if any difficulty is encountered if this is done within 24–48 h from the time the old catheter was lost. In cases where the guidewire passes to the venous entry site, but will not make the curve to advance downward into the venous system, a vascular catheter such as Kumpe can be used to both stiffen the guidewire and redirect its tip.

Anecdotal experience with this procedure on multiple occasions has shown it to not only be effective but also safe.

Fig. 36.6 New catheter in place. *A* Catheter

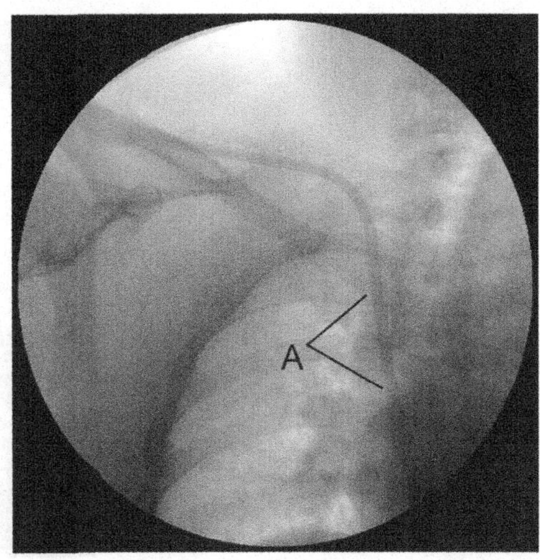

Fig. 36.5 Guidewire in place, ready for catheter insertion. *A* Guidewire

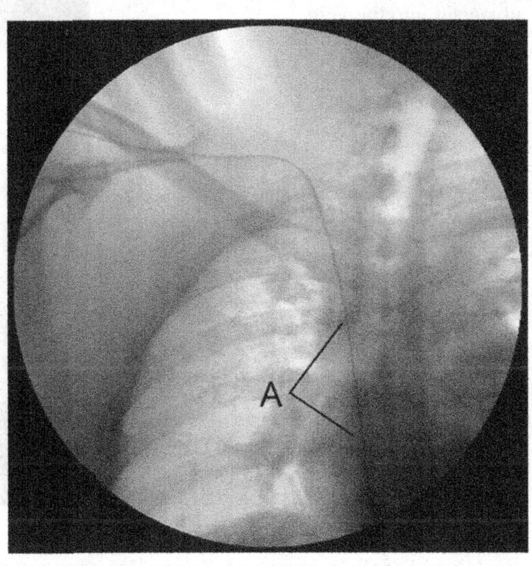

Chapter 37
Internal Jugular Vein Catheter Inserted Through Side of Stent

Gerald A. Beathard

Case Presentation

The patient was a 62-year-old male who had been on dialysis for 12 years. The patient had multiple failed peripheral arteriovenous accesses and had become catheter dependent. He was dialyzing with a left internal jugular tunneled cuffed catheter which was thrombosed. His history indicated that this catheter had been exchanged 6 times in the past 5 months. He was referred to the dialysis access clinic for a new catheter.

Examination of the patient revealed that he had multiple old abandoned peripheral arteriovenous accesses and had multiple scars over his anterior chest at the sites of the previous tunneled catheter placements. With ultrasound he was found to have a patent right internal jugular vein; however, it was known that this vein had been "jailed" with a stent which had been placed in the right subclavian vein, extending into the brachiocephalic vein.

After the patient was prepped and draped, the right internal jugular vein was cannulated with a micropuncture needle (Fig. 37.1). The 0.018-inch guidewire was inserted and advanced through the side of the stent. The insertion of a micropuncture dilator was followed by exchanging the microguidewire for a 0.035-inch standard guidewire (Fig. 37.2), which was passed down into the inferior vena cava. The new dialysis catheter was then placed using the standard technique without any complications (Fig. 37.3). Since a Tesio catheter was inserted, this process was performed two times. Once the new catheter was in place, the old thrombosed left internal jugular catheter was removed and discarded.

The new catheter flushed easily with a 10-mL syringe. The patient was sent back to the dialysis clinic and dialyzes without difficulty.

G.A. Beathard, MD, PhD (✉)
Lifetime Vascular Access, University of Texas Medical Branch,
5135 Holly Terrace, Houston, TX 77056, USA
e-mail: gbeathard@msn.com

© Springer International Publishing AG 2017
A.S. Yevzlin et al. (eds.), *Dialysis Access Cases*,
DOI 10.1007/978-3-319-57500-1_37

Fig. 37.1 Fluoroscopic appearance of chest prior to beginning procedure. *A* Stent in right subclavian/brachiocephalic vein jailing right internal jugular vein

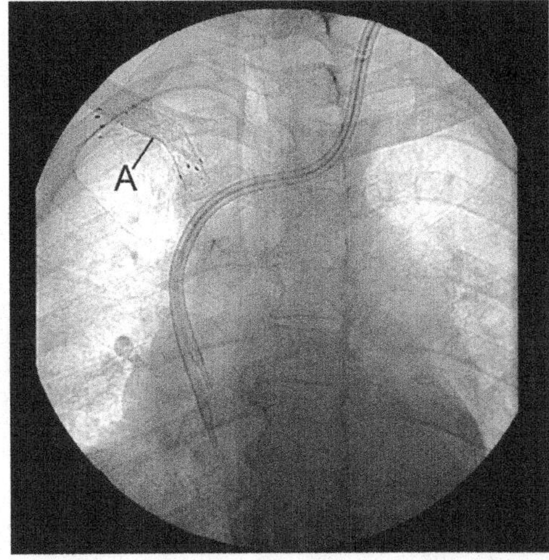

Fig. 37.2 Cannulation passing through stent. *A* Dilator passing through stent, *B* tip of dilator in superior vena cava

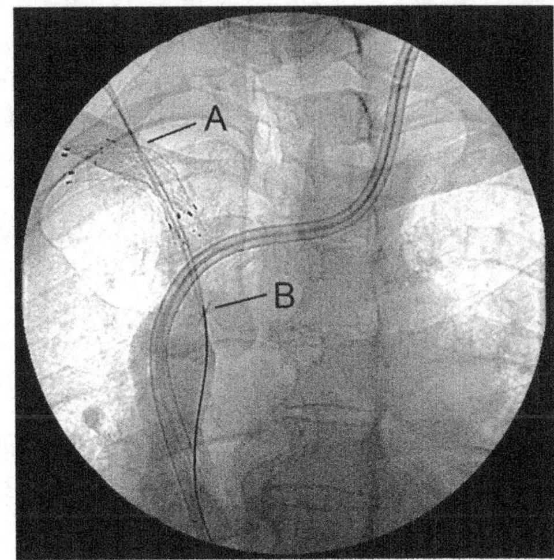

Discussion

One of the basic tenets of stent placement is to avoid jailing major veins; however, the junction between the subclavian and brachiocephalic veins is a common site for stenosis to occur associated with the use of a central venous catheter. This being the case, it is not unusual for a stent to be placed at this location. In these cases, if the insertion of a dialysis catheter is required at a later date, inserting the catheter through the side of the stent, as in this case, becomes necessary. Although one has to consider the possibility of a catheter-related bloodstream infection infecting the stent secondarily, such an occurrence has not been documented in the literature.

Fig. 37.3 New catheter in place passing through stent, old catheter not yet removed

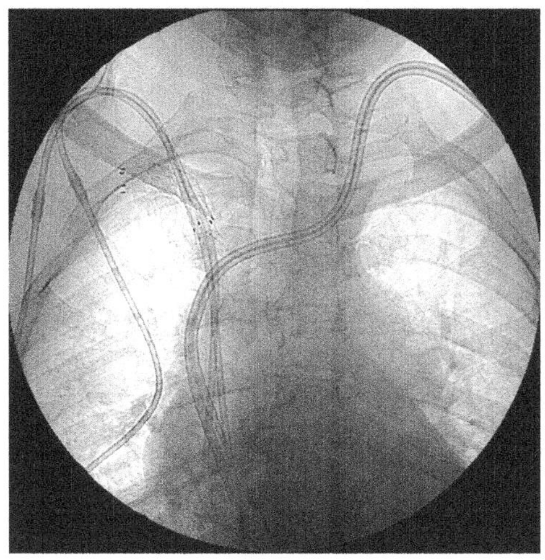

Chapter 38
Pneumothorax Occurring with Dialysis Catheter Insertion

Gerald A. Beathard

Case Presentation

The patient was a 63-year-old male with end-stage renal disease who was referred to the dialysis access center for the placement of a tunneled dialysis catheter in order to initiate dialysis. After the patient was prepped and draped, the right internal jugular vein was cannulated under direct ultrasound guidance. A guidewire was inserted. After the catheter had been tunneled from the exit site up to the venotomy site, graded dilators were used over the guidewire in preparation for catheter insertion. When passing the first dilator, resistance was encountered. The operator continued to advance the dilator at which time the patient began to complain of shortness of breath. The procedure was stopped with the dilator still in place. The patient's oxygen saturation dropped to 82% and he became bradycardic. At this time, fluoroscopy of the patient's chest revealed a large pneumothorax (Fig. 38.1). Because of the patient symptoms, it was obvious that he had a tension pneumothorax. A chest tube was inserted using a pneumothorax set (Arrow International, Pennsylvania). The patient's condition began to improve immediately and he was transferred to the hospital.

Discussion

Unless the subclavian vein is being cannulated, pneumothorax (PT) as a complication for the placement of a dialysis catheter is not a common event [1]. In a report of 1765 tunneled dialysis catheters placed in the internal jugular vein by

G.A. Beathard, MD, PhD (✉)
Lifetime Vascular Access, University of Texas Medical Branch,
5135 Holly Terrace, Houston, TX 77056, USA
e-mail: gbeathard@msn.com

© Springer International Publishing AG 2017
A.S. Yevzlin et al. (eds.), *Dialysis Access Cases*,
DOI 10.1007/978-3-319-57500-1_38

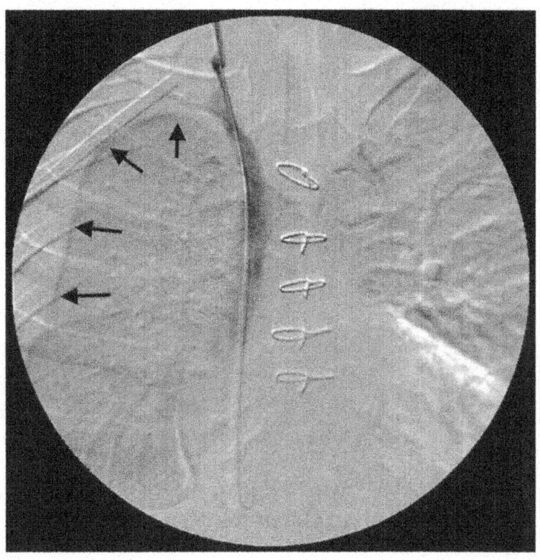

Fig. 38.1 Appearance of chest fluoroscopy showing large right-sided pneumothorax. *Arrows* indicate the edge of the right lung

nephrologists using ultrasound [2], there was only a single case of PT. A prospective study of 450 cases using ultrasound guidance and internal jugular placement encountered no cases [3]. However, when that rare case does occur, early recognition and appropriate management can be critical [4, 5]. PT can be divided into two categories – asymptomatic and symptomatic. The management of these is quite different.

The symptoms produced by a PT are dependent primarily upon its size. If the volume of air that enters the pleural space is small, there may be no associated symptoms. The condition is recognized only radiographically, if at all. A multicenter, prospective, observational study was conducted that reported on more than 500 trauma patients with asymptomatic PT identified on CT scan, with an initially normal chest radiograph. The study arms included observation versus chest tube thoracostomy. Only 6% of patients failed observation. This failure was seen only in patients with chest radiographic evidence of PT progression and symptoms of respiratory distress. According to this study, it is safe to closely observe patients with asymptomatic PT noted only on chest radiographs [6]. This should be done in the hospital setting, however. Oxygen administration at 3 L/min nasal canula or higher in these cases can be used to treat possible hypoxemia (not generally present) and has been reported to be associated with a fourfold increase in the rate of pleural air resorption compared with room air alone [6].

If the volume of air in the pleural space is large as in this case, symptoms occur. Immediate recognition and appropriate management is critical. In most instances one is dealing with a tension pneumothorax (TPT). This is defined as the accumulation of air under pressure within the pleural space. With inspiration, air enters the pleural space, but with expiration it cannot exit. Symptoms associated with this situation are variable but can be dramatic and tend to get progressively worse. The increasing volume of air in the pleural space compresses the lung, displaces the mediastinum and its structures toward the opposite side, and eventually causes cardiopulmonary impairment [7]. Rapid deterioration will occur if remedial action is not taken promptly.

When the clinical situation is such that a TPT is suspected, the first step is to immediately evaluate the ABC mnemonic (airway, breathing, and circulation). Be prepared to support these functions. Administer oxygen, ventilate the patient, and establish an intravenous line if one is not already available (the dialysis catheter may serve if it has been appropriately placed). Immediate thoracostomy must be performed in any patient who presents with hemodynamic instability or hypoxia [4, 5]. Since immediate, effective treatment may be necessary to save the patient's life, it is essential that the interventional facility have the equipment and supply items necessary to manage a TPT until the patient can be transported to the hospital. Failure can result in rapid clinical deterioration and cardiac arrest.

There are several ways that an effective thoracostomy can be accomplished using commercially available kits such as the pneumothorax set (Arrow International, Pennsylvania) and the thoracic vent (Tru-Close®, Uresil, Skokie, IL). If one of these is not available, the insertion of large bore needle or catheter may be lifesaving as an emergency measure. Needle insertion is most safely accomplished from a lateral approach at the fifth intercostal space at the anterior axillary line [8–11]. After entering the pleural space (indicated by aspiration of air), the needle can be connected to a three-way stopcock and a large syringe can be used to aspirate the air. Manual aspiration should be continued until no more air could be aspirated.

In a study involving manual aspiration [12], 102 cases of PT following interventional radiological procedures underwent percutaneous manual aspiration of a PT. Air was aspirated from the pleural space using an 18- or 20-gauge intravenous catheter attached to a three-way stopcock and 20- or 50-mL syringe. In 87 of the 102 patients (85.3%), the pneumothorax had resolved completely on follow-up chest radiographs without chest tube placement. This success rate was subsequently confirmed in a larger series of 243 cases by the same investigators [13].

References

1. Peris A, Zagli G, Bonizzoli M, et al. Implantation of 3951 long-term central venous catheters: performances, risk analysis, and patient comfort after ultrasound-guidance introduction. Anesth Analg. 2010;111:1194.
2. Beathard GA, Litchfield T. Effectiveness and safety of dialysis vascular access procedures performed by interventional nephrologists. Kidney Int. 2004;66:1622.
3. Karakitsos D, Labropoulos N, De Groot E, et al. Real-time ultrasound-guided catheterisation of the internal jugular vein: a prospective comparison with the landmark technique in critical care patients. Crit Care. 2006;10:R162.
4. American College of Surgeons, Committee on Trauma. Advanced trauma life support course manual. Chicago: American College of Surgeons; 2004.
5. McPherson JJ, Feigin DS, Bellamy RF. Prevalence of tension pneumothorax in fatally wounded combat casualties. J Trauma. 2006;60:573.
6. Moore FO, Goslar PW, Coimbra R, et al. Blunt traumatic occult pneumothorax: is observation safe?--results of a prospective, AAST multicenter study. J Trauma. 2011;70:1019.

7. Kaufmann C. Initial assessment and management. New York: McGraw-Hill Medical; 2008.
8. Rawlins R, Brown KM, Carr CS, Cameron CR. Life threatening haemorrhage after anterior needle aspiration of pneumothoraces. A role for lateral needle aspiration in emergency decompression of spontaneous pneumothorax. Emerg Med J. 2003;20:383.
9. Heng K, Bystrzycki A, Fitzgerald M, et al. Complications of intercostal catheter insertion using EMST techniques for chest trauma. ANZ J Surg. 2004;74:420.
10. Jenkins C, Sudheer PS. Needle thoracocentesis fails to diagnose a large pneumothorax. Anaesthesia. 2000;55:925.
11. Biffl WL. Needle thoracostomy: a cautionary note. Acad Emerg Med. 2004;11:795.
12. Yamagami T, Kato T, Hirota T, et al. Usefulness and limitation of manual aspiration immediately after pneumothorax complicating interventional radiological procedures with the transthoracic approach. Cardiovasc Intervent Radiol. 2006;29:1027.
13. Yamagami T, Terayama K, Yoshimatsu R, et al. Role of manual aspiration in treating pneumothorax after computed tomography-guided lung biopsy. Acta Radiol. 2009;50:1126.

Chapter 39
Perforation of Superior Vena Cava with Dialysis Catheter Insertion

Gerald A. Beathard

Case Presentation

The patient was a 72-year-old male who was referred to the dialysis access center for the placement of a tunneled dialysis catheter. The patient was scheduled to have an arteriovenous fistula placed in the right arm. For this reason, the left internal jugular vein was cannulated after the patient had been prepped and draped. The vein was cannulated without difficulty and a guidewire was inserted. Following the use of graded dilators, a 32 cm step-tip cuffed dialysis catheter was inserted through a sheath over the guidewire (Fig. 39.1). Examination of the patient's thorax during the insertion of the catheter revealed that the guidewire appeared to extend beyond the limits of the mediastinum (Fig. 39.2). At this time the guidewire was removed (Fig. 39.3) and radiocontrast was injected through the catheter. This showed an accumulation of radiocontrast in the pleural space (Fig. 39.4). At this point, the procedure was discontinued and the patient was sent to the hospital with the catheter in place. At the hospital the catheter was removed by vascular surgery without significant consequences. The patient was observed for 24 h, was doing well, and was discharged.

Discussion

Sequelae resulting from acute perforation of a central vein can be quite variable. Many interventionalists have had the experience of perforating a central vein and observing no adverse effect. This may be due in part to the rather low pressure

G.A. Beathard, MD, PhD (✉)
Lifetime Vascular Access, University of Texas Medical Branch,
5135 Holly Terrace, Houston, TX 77056, USA
e-mail: gbeathard@msn.com

© Springer International Publishing AG 2017
A.S. Yevzlin et al. (eds.), *Dialysis Access Cases*,
DOI 10.1007/978-3-319-57500-1_39

Fig. 39.1 Beginning to insert catheter over the guidewire. *A* Guidewire entering the superior vena cava

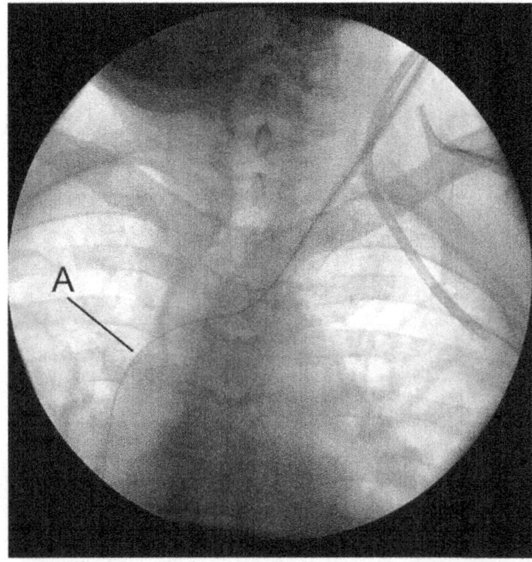

Fig. 39.2 Appearance of dialysis catheter outside limits of mediastinum. *A* Tip of catheter

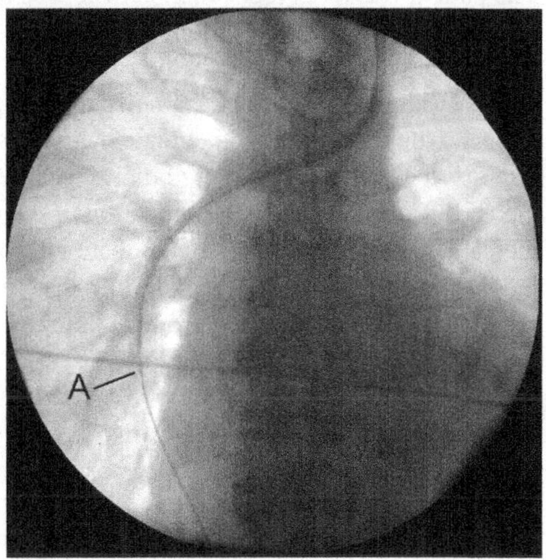

within the right atrium and superior vena cava (2–6 mmHg normally and even lower if patient is hypovolemic) and the fact that the perforation was high in the superior vena cava (above the pericardial reflection) or in the brachiocephalic vein. However, perforation can lead to bleeding either into the pleural space, into the mediastinum, or into the pericardium [1–6]. The upper half of the superior vena cava is covered by mediastinal connective tissue. In this region perforation can result in hemo- or hydrothorax/mediastinum [7]. Life-threatening cardiac tamponade may result if a catheter perforates the right heart or that part of the superior vena cava that is covered by pericardium (pericardial reflection).

With large-bore hemodialysis catheters, injuries may cause severe bleeding leading to hemodynamic instability, hypovolemic shock, and even death [2, 3, 8, 9].

Fig. 39.3 Appearance of
dialysis catheter with
guidewire removed. *A* Tip
of catheter

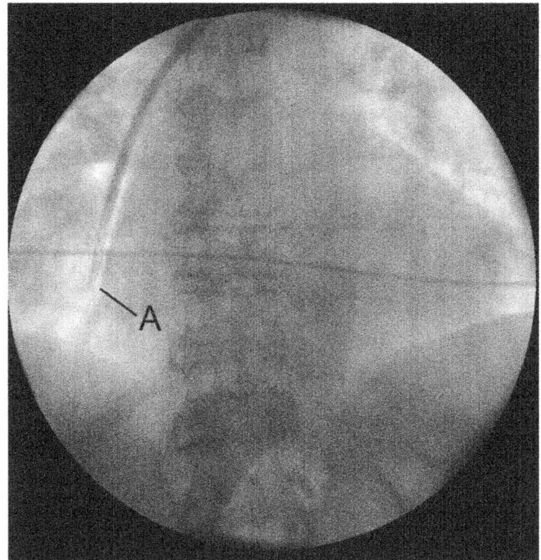

Fig. 39.4 Presence of
radiocontrast in pleural
space following injection
through catheter. *A* Tip of
catheter

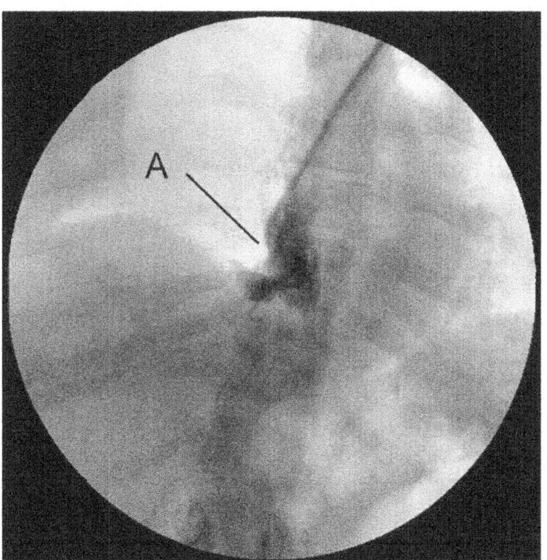

Serious damage to the central veins can result from failure to use the larger-bore venous dilators properly when preparing a cannulation site for the insertion of the dialysis catheter. To be safe, the dilator must follow the course of the guidewire into and down the central vein. This path is not completely straight. Initially, it angles downward as one enters the vein and then turns caudally following the path of the central vein (much more circuitous than this on the left). Additionally, because of the close proximity of the patient's neck and head to the entry site, there is a tendency for the operator to angle the peripheral end of the dilator outward.

Because of these issues, the guidewire can become kinked, and instead of serving as a track to direct the passage of the device, it is advanced along with it. If the tip of the dilator is pressing against the side of the vein, the advancing guidewire can act as a blade, slicing the vein as it moves along its wall [3]. Some force is frequently required with large-bore dilator insertion to pass the device through the superficial structures and enter the vein. If the guidewire is kinked, this force may be applied toward destruction of the vein. The result can be disastrous. Actually, for a successful catheter insertion, it is necessary for the dilator to only be passed far enough into the vein to dilate the entry site. In many instances, the operator advances it considerable beyond this point making this problem, when it occurs, much worse.

In a retrospective review of 4000 cases of catheter placement [10], 10 patients had a vascular perforation, an incidence of 0.25%. Operator error was the primary cause of the problem; the injury was caused by kinking of the guidewire followed by forcing the vessel dilator or peel-away sheath into the central vein. The initial error, kinking of the guide wire, was the result of the operator's failure to firmly hold and stabilize the guide wire while advancing the vessel dilator. Of these ten cases, four were fatal.

This problem can be prevented by checking for free passage of the guidewire in and out of the dilator repeatedly during insertion as described above. Additionally, direct observation under fluoroscopy will alert the operator to any untoward dilator-guidewire interaction. Careful attention to these two issues will prevent the occurrence of this problem.

Since the effects of central vein perforation are variable, the management will also be variable and largely dependent upon the observation of changes in the individual patient. The sequelae of perforation are largely dependent upon the size of the device involved and whether the perforation is into the pleural space or into the mediastinum. If it is only the needle, there are usually no ill effects and no treatment is required. One may not even recognize that it has occurred unless the guidewire is passed into the pleural space. Even then, withdrawal is generally not followed by adverse changes.

With a larger device such as a dilator or catheter, management will depend on the extent of the injury, general status of the patient, and efficiency of the patient's clotting system [4]. It should be remembered that the removal of the offending device can result in massive hemorrhage requiring surgical intervention which cannot be accomplished in a free-standing facility. Passage into the mediastinum may result in a hematoma that may be self-limiting. Passage into the pleural space can result in a hemothorax (or a pneumothorax, if the device is open to the air). If the event results in a large defect, it is frequently fatal.

One can never be sure what will happen once the device is removed leaving a large defect in the vein. It is possible that massive hemorrhage requiring an exploratory thoracotomy to repair the defect can occur [8, 11]. It is better to leave the device in place and transfer the patient to a hospital setting (in cases down in an outpatient environment) where it can be removed with surgical support available.

There are reports of endovascular management of such an injury [4, 12]. In one case report [12], the perforation site was repaired using a stent graft. In this instance the stent graft was placed through a femoral approach prior to withdrawal of the catheter.

References

1. Wicky S, Meuwly JY, Doenz F, Uske A, Schnyder P, Denys A. Life-threatening vascular complications after central venous catheter placement. Eur Radiol. 2002;12(4):901–7.
2. Schummer W, Schummer C, Fritz H. Perforation of the superior vena cava due to unrecognized stenosis. Case report of a lethal complication of central venous catheterization. Anaesthesist. 2001;50(10):772–7.
3. Jankovic Z, Boon A, Prasad R. Fatal haemothorax following large-bore percutaneous cannulation before liver transplantation. Br J Anaesth. 2005;95(4):472–6.
4. Wadelek J, Drobinski D, Szewczyk P, Abbas F, Franczyk M, Niewinska M, et al. Perforation of the internal jugular vein during cannulation for haemodialysis. Anestezjol Intens Ter. 2009;41(2):110–3.
5. Jost K, Leithauser M, Grosse-Thie C, Bartolomaeus A, Hilgendorf I, Andree H, et al. Perforation of the superior vena cava - a rare complication of central venous catheters. Onkologie. 2008;31(5):262–4.
6. Ko SF, Ng SH, Fang FM, Wan YL, Hsieh MJ, Liu PP, et al. Left brachiocephalic vein perforation: computed tomographic features and treatment considerations. Am J Emerg Med. 2007;25(9):1051–6.
7. Bayer O, Schummer C, Richter K, Frober R, Schummer W. Implication of the anatomy of the pericardial reflection on positioning of central venous catheters. J Cardiothorac Vasc Anesth. 2006;20(6):777–80.
8. Wang CY, Liu K, Chia YY, Chen CH. Bedside ultrasonic detection of massive hemothorax due to superior vena cava perforation after hemodialysis catheter insertion. Acta Anaesthesiol Taiwanica. 2009;47(2):95–8.
9. Elwali A, Drissi M, Bensghir M, Ahtil R, Azendour H, Kamili ND. Fatal evolution of a left brachio-cephalic vein perforation by a catheter for hemodialysis. Nephrol Ther. 2011;7(7):562–4.
10. Robinson JF, Robinson WA, Cohn A, Garg K, Armstrong 2nd JD. Perforation of the great vessels during central venous line placement. Arch Intern Med. 1995;155(11):1225–8.
11. Florescu MC, Mousa A, Salifu M, Friedman EA. Accidental extravascular insertion of a subclavian hemodialysis catheter is signaled by nonvisualization of catheter tip. Hemodial Int. 2005;9(4):341–3.
12. Azizzadeh A, Pham MT, Estrera AL, Coogan SM, Safi HJ. Endovascular repair of an iatrogenic superior vena caval injury: a case report. J Vasc Surg. 2007;46(3):569–71.

Part IV
Draining Veins

Chapter 40
Grade II Extravasation Complicating Venous Angioplasty

Gerald A. Beathard

Case Presentation

The patient was a 28-year-old female hemodialysis patient with a forearm loop graft in her left arm, who was referred to the dialysis access center because of poor access blood flow measured on dialysis. The examination of the loop graft revealed that it was hyper-pulsatile and there was a localized thrill at the venous anastomosis which extended up the draining vein for approximately 6 cm. The bruit heard over this area was high pitched with a shortened diastolic component.

After the patient's arm was prepped and draped, it was cannulated and a sheath was put in place. An angiogram was performed which showed an area of marked stenosis in the draining vein approximately 6 cm above the venous anastomosis (Fig. 40.1, arrow). After a guidewire was passed, an 8 × 4 angioplasty balloon was put in place in the lesion which was dilated. The patient complained of continuing pain in the area treated after balloon deflation and swelling was noted in the area. The access inflow was immediately included manually. A post-angioplasty angiogram was performed which showed extravasation of radiocontrast with no flow going past the area of extravasation (Fig. 40.2). The extravasation was stable and not continuing to enlarge. The angioplasty balloon was reinserted and positioned in the area of extravasation. A low-pressure inflation was performed to tapenade the area (Fig. 40.3) while the access inflow continued to be manually occluded. After the balloon had been in place for 5 min, it was deflated and removed. A follow-up angiogram showed good flow through the area (Fig. 40.4). Hemostasis was obtained and the patient was returned to the dialysis facility.

G.A. Beathard, MD, PhD (✉)
Lifetime Vascular Access, University of Texas Medical Branch,
5135 Holly Terrace, Houston, TX 77056, USA
e-mail: gbeathard@msn.com

© Springer International Publishing AG 2017
A.S. Yevzlin et al. (eds.), *Dialysis Access Cases*,
DOI 10.1007/978-3-319-57500-1_40

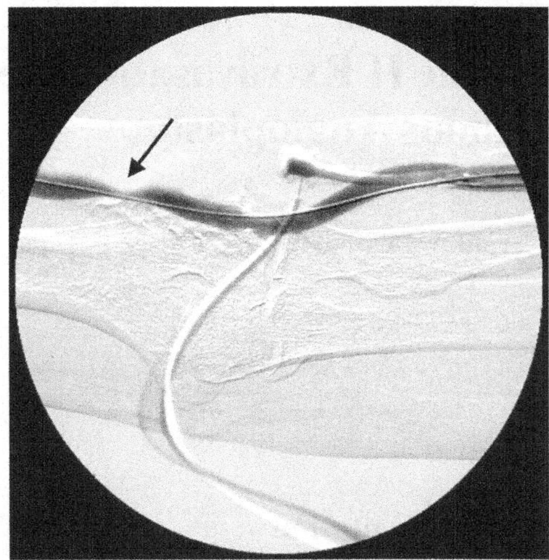

Fig. 40.1 Angiogram showing appearance of stenotic lesion (*arrow*)

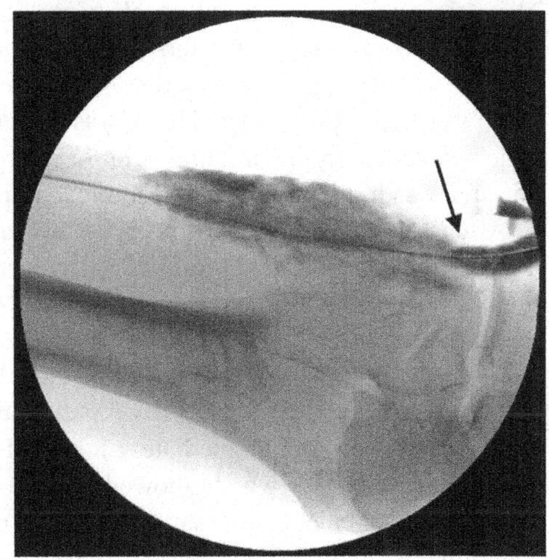

Fig. 40.2 Appearance of extravasation of radiocontrast. The *arrow* indicates the point at which the column of radiocontrast stops due to the obstruction created by the extravasation

Discussion

Venous stenosis occurring in association with a dialysis vascular access is the result of neointimal hyperplasia. It is obvious that this lesion has tremendous tensile strength when one considers that some lesions require more than 20 atm of pressure to treat [1]. For an angioplasty to be successful, it is probable that some degree of mechanical injury to the vessel wall must occur [2, 3]. However, a deep tear or rupture through the vein wall can occur. This can be fluoroscopically identified as extravasation of contrast material into the perivascular tissues.

The severity and consequences of venous rupture resulting in extravasation vary from minimal to severe. A classification system for venous extravasation based

Fig. 40.3 Balloon tapenade of the point of extravasation

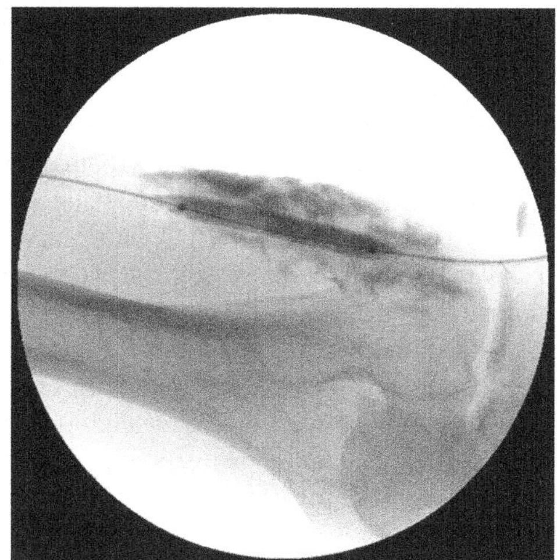

Fig. 40.4 Post-tapenade angiogram showing restoration of flow through graft and draining veins

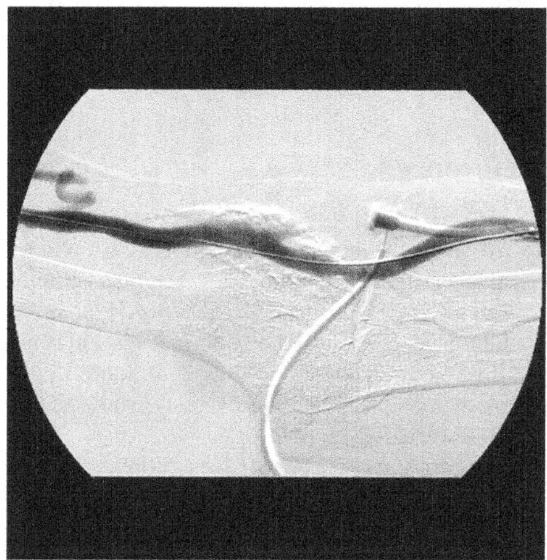

upon the clinical features demonstrated has been proposed [4, 5]. Basically, Grade I extravasation is characterized by being stable and not affecting access blood flow; Grade II extravasation is stable but either slows or stops access blood flow; Grade III extravasation is unstable with continuing, active extravasation.

Although angioplasty-induced venous rupture is the most common complication associated with angioplasty [6], it does not occur with the high frequency. In a review of 305,470 angioplasty cases (unpublished data), venous rupture with extravasation occurred in 1.38% of fistulas and 0.66% of grafts. The majority of these were Grade I. The incidence of Grade II extravasation was 0.27% in fistulas and 0.14% in grafts.

When extravasation is noted, one must make two determinations – (1) is it stable or continuing to enlarge, and (2) does it affect flow. As soon as it is obvious that a vascular rupture has occurred and extravasation is occurring, the access should be manually occluded upstream (distal) to the site of rupture to relieve pressure on the site and limit the degree of extravasation until a determination of its severity can be determined. In this case described here, a Grade II extravasation was apparent.

Blood flow will move toward the path of least resistance. As long as the lumen is not obstructed, this will be the route that it will take. However, if there is obstruction such as a tear in the vessel wall creating an obstructing flap as well as an alternative pathway (the hole in the vessel wall), extravasation occurs. In addition to the intra-luminal flap, the external pressure exerted by the extravasation on the vessel tends to occlude the lumen. The goal of treatment is to press the flap outward and com-press the extravascular accumulation in order to open the lumen and restore flow. Once the lumen is opened with an angioplasty balloon, the sustained pressure of the balloon (tamponade) tends to stabilize the situation after a few minutes and allow for the restoration of flow. This return of flow helps to keep the offending flap in position.

Not all cases of Grade II extravasation can be treated with balloon tapenade. If this is tried and failed to work the first time, it should be tried again. If this is unsuc-cessful, the placement of an endovascular stent will generally restore flow.

References

1. Vesely TM, Pilgram TK. Angioplasty balloon inflation pressures during treatment of hemodi-alysis graft-related stenoses. J Vasc Interv Radiol. 2006;17:623.
2. Davidson CJ, Newman GE, Sheikh KH, et al. Mechanisms of angioplasty in hemodialysis fis-tula stenoses evaluated by intravascular ultrasound. Kidney Int. 1991;40:91.
3. Higuchi T, Okuda N, Aoki K, et al. Intravascular ultrasound imaging before and after angio-plasty for stenosis of arteriovenous fistulae in haemodialysis patients. Nephrol Dial Transplant. 2001;16:151.
4. Beathard GA. Angioplasty for arteriovenous grafts and fistulae. Semin Nephrol. 2002;22:202.
5. Beathard GA. Management of complications of endovascular dialysis access procedures. Semin Dial. 2003;16:309.
6. Kornfield ZN, Kwak A, Soulen MC, et al. Incidence and management of percutaneous translu-minal angioplasty-induced venous rupture in the "fistula first" era. J Vasc Interv Radiol. 2009;20:744.

Chapter 41
Grade III Venous Rupture

Gerald A. Beathard

Case Presentation

The patient was a 54-year-old male who had been on dialysis for 3 years. The patient had a left brachial-cephalic arteriovenous fistula (AVF). He was referred to the dialysis vascular access facility because of poor blood flow detected at the dialysis clinic. Physical examination revealed that the AVF was hyperpulsatile. A thrill was palpable over the AVF just below the shoulder.

After the patient was prepped and draped, an angiogram was performed which showed a tight stenosis in the cephalic vein (Fig. 41.1). A high-pressure 8 × 4 angioplasty balloon was used first in an attempt to dilate the lesion. This was unsuccessful. An ultrahigh-pressure 8 × 4 angioplasty balloon was then used. The lesion opened with a distinct palpable "pop" and full balloon effacement was observed (Fig. 41.2). Although the patient was sedated, he appeared to be restless and uncomfortable. The area of the lesion was palpated which revealed an enlarging, pulsatile hematoma. AVF inflow was immediately interrupted with manual compression just above the anastomosis. An angiogram revealed that there was no flow beyond the site of extravasation (Fig. 41.3). Balloon tapenade was attempted twice without any effect. The hematoma continued to be pulsatile when the inflow occlusion was released. A decision was made to place a stent-graft. An 8 × 8 stent-graft was inserted without difficulty (Fig. 41.4). This restored blood flow with no further extravasation (Fig. 41.5).

G.A. Beathard, MD, PhD (✉)
Lifetime Vascular Access, University of Texas Medical Branch,
5135 Holly Terrace, Houston, TX 77056, USA
e-mail: gbeathard@msn.com

© Springer International Publishing AG 2017
A.S. Yevzlin et al. (eds.), *Dialysis Access Cases*,
DOI 10.1007/978-3-319-57500-1_41

Fig. 41.1 Initial angiogram. *A* Stenotic lesion

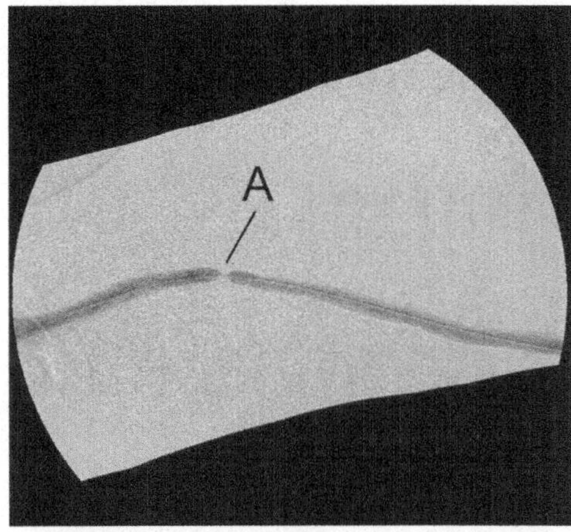

Fig. 41.2 Angioplasty of lesion showing full effacement

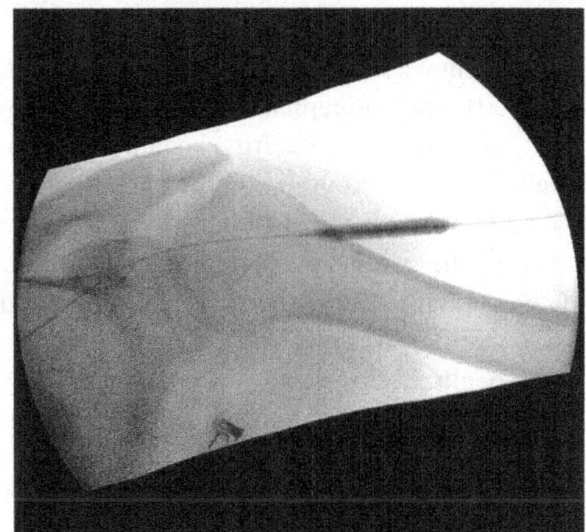

Discussion

The most frequent procedure-related complication seen in association with angioplasty that dictates the need for intervention is vein rupture with extravasation [1]; however, this is not a frequent occurrence. The clinical significance of vein rupture with extravasation and hematoma formation is variable, ranging from none to disaster for the access. The difference lies in the severity of the tear. The amount of extravasated radiocontrast associated with the hematoma may be minimal or absent. In actuality the size of the hematoma is not particularly important in determining the sequelae with which it is associated. One may see a small degree of extravasation with hematoma formation as a result of complete vein disruption that was quickly controlled. On the other hand, a very large hematoma may be completely stable and have no effect on blood flow.

Fig. 41.3 Extravasation at angioplasty site. *A* Point of blockage of blood flow

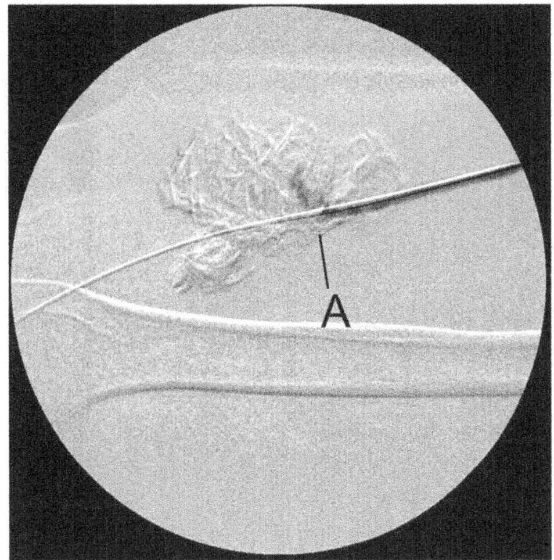

Fig. 41.4 Stent graft in place. *A* Stent graft

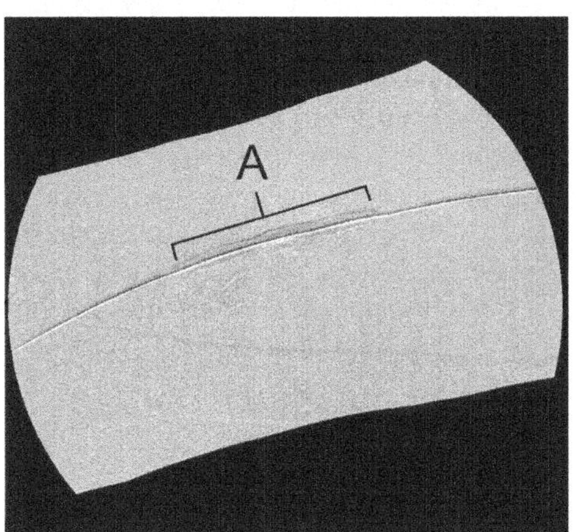

As soon as it is obvious that a vascular rupture has occurred and a hematoma is forming, the access should be manually occluded upstream (distal) to the site of rupture to relieve pressure on the site and limit the degree of extravasation until a determination of its severity can be determined. The severity of the situation is determined by answering two questions – (1) is it stable or contorting to a large hematoma, and (2) does it affect blood flow? A classification system for extravasation has been devised based upon these two issues [2–4].

- Grade 1 – does not interfere with blood flow, stable
- Grade 2 – blood flow is slowed or stopped, stable
- Grade 3 – unstable, progressively enlarging hematoma

With the grade 3 hematoma, there is continuing active bleeding from a disrupted vessel. Arterial blood is being pumped directly into the surrounding tissue.

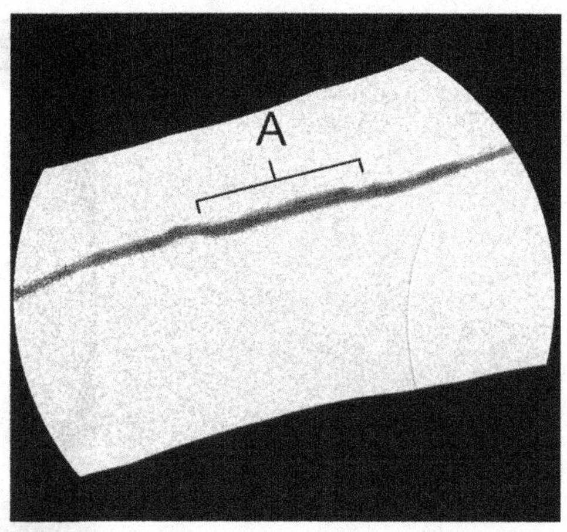

Fig. 41.5 Postprocedure angiogram showing flow through stent graft. *A* Stent graft area

Hematoma formation occurs very rapidly, the size of which depends on how quickly the condition is recognized and controlled. With the development of this complication, there is a definite risk for losing the access. The first goal in management is to control the extravasation. This can generally be done by simply manually occluding the inflow to the access. The second goal is to attempt to salvage the access. In accomplishing this it is critical that the guidewire be left in place. If it has been inadvertently removed, replacing it is extremely difficult and often impossible. Salvage can generally be accomplished by the placement of a stent-graft. If this is not effective, the access will need to be thrombosed.

References

1. Beathard GA, Litchfield T, Physician Operators Forum of RMS Lifeline Inc. Effectiveness and safety of dialysis vascular access procedures performed by interventional nephrologists. Kidney Int. 2004;66:1622.
2. Beathard GA. Angioplasty for arteriovenous grafts and fistulae. Semin Nephrol. 2002;22:202.
3. Beathard GA. Management of complications of endovascular dialysis access procedures. Semin Dial. 2003;16:309.
4. Beathard GA, Urbanes A, Litchfield T. The classification of procedure-related complications – a fresh approach. Semin Dial. 2006;19:527.

Chapter 42
Parallel Guidewire Angioplasty

Gerald A. Beathard

Case Presentation

The patient was a 42-year-old male with a left brachial-basilic arteriovenous fistula (AVF) who had been on dialysis for 2 years. The patient was referred to the vascular access center because of low blood flow on dialysis and a hyperpulsatile fistula. Upon examination the patient's AVF was found to be dilated and hyperpulsatile with a prominent thrill and bruit at the level of the angle of transposition (swing point lesion).

After the patient was prepped and draped, an angiogram was performed. This showed an area of marked stenosis at the angle of transposition of the AVF (Fig. 42.1). The remainder of the angiogram was normal. Using a high-pressure 10 × 4 angioplasty balloon, an attempt was made to dilate the lesion. This was unsuccessful and was discontinued because the angioplasty balloon appeared to be on the verge of rupture (Fig. 42.2). It was replaced with an ultrahigh-pressure 10 × 4 angioplasty balloon and another attempt was made to dilate the lesion. This was also unsuccessful (Fig. 42.3); the attempt ended with rupture of the angioplasty balloon.

At this time, a second guidewire was introduced through the sheath and advanced up to the level of the central veins. The sheath was removed and reintroduced over only one of the guidewiresa, excluding the second one. Another ultrahigh-pressure 10 × 4 angioplasty balloon was introduced and advanced up to the site of the lesion. The angioplasty balloon was then inflated to 30 atm forcing the parallel guidewire against the wall of the vessel (Fig. 42.4). This resulted in full effacement of angioplasty balloon (Fig. 42.5). A post-procedure angiogram showed full dilatation of the lesion with no residual (Fig. 42.6).

G.A. Beathard, MD, PhD (✉)
Lifetime Vascular Access, University of Texas Medical Branch,
5135 Holly Terrace, Houston, TX 77056, USA
e-mail: gbeathard@msn.com

© Springer International Publishing AG 2017
A.S. Yevzlin et al. (eds.), *Dialysis Access Cases*,
DOI 10.1007/978-3-319-57500-1_42

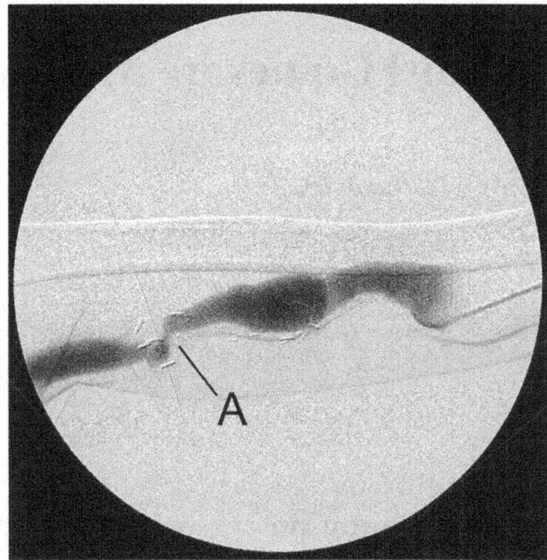

Fig. 42.1 Angiogram of stenotic lesion at swing point in brachial-basilic AVF. *A* Initial lesion

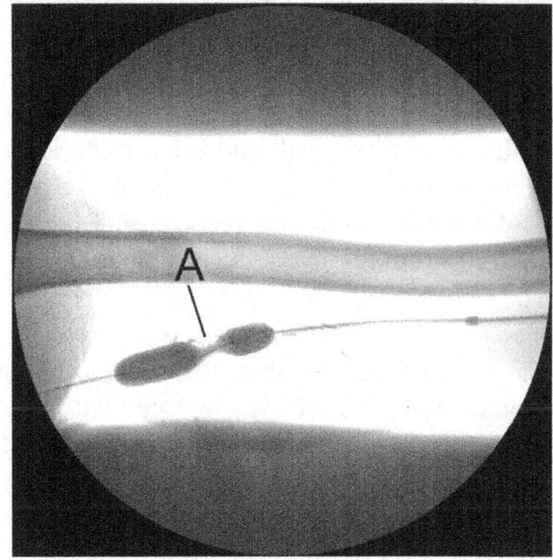

Fig. 42.2 Angioplasty with ultrahigh-pressure balloon (note rounded ends of balloon). *A* Site of resistance lesion

Discussion

Some lesions are extremely resistant to dilatation. When this occurs there are three courses of action that are available – more pressure, do something to weaken the lesion or refer for surgical revision. The simplest of these approaches is the application of more pressure. The standard high-pressure angioplasty balloon is designed to allow for 16–20 atm of pressure. If this is exceeded, there is a risk of balloon rupture. With some angioplasty balloons, one can attack an imminent rupture by watching the configuration of the balloon. Just before the balloon ruptures, it is normally tapered; ends will frequently take on a rounded configuration (Fig. 42.2).

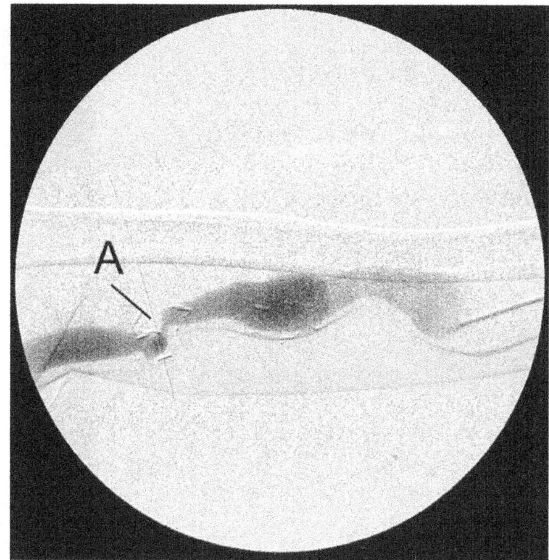

Fig. 42.3 Post-angioplasty angiogram following balloon rupture. *A* Site of resistant lesion

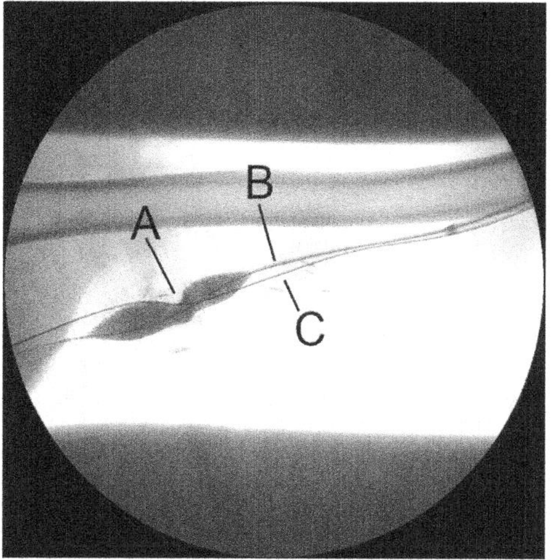

Fig. 42.4 Repeat angioplasty using cutting wire. *A* Site of resistant lesion, *B* angioplasty balloon guidewire, *C* second guidewire being used as cutting wire

If full effacement is not accomplished using a standard high-pressure balloon, the next step to achieve the application of greater dilating force is to move to an ultrahigh-pressure balloon. This device will tolerate more than 30 atm of pressure, and with its use, it is possible to treat most of resistant lesions successfully. In a study [1] that included 230 lesions that were treated with a goal of total angioplasty balloon effacement, it was found that only 1% of the lesions could not be successfully treated. A total of 55% of lesions required pressures greater than 15 atm to obtain complete effacement. When initial failures were excluded, it was found that 20% of the lesions in AVFs and 9% in arteriovenous graft (AVG) required pressures greater than 20 atm to obtain complete effacement.

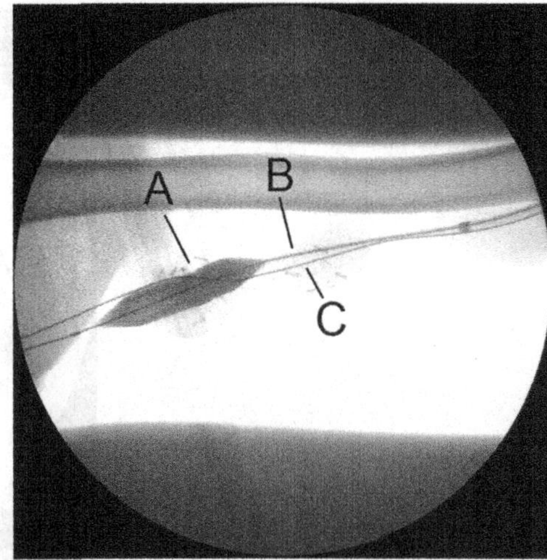

Fig. 42.5 Total angioplasty balloon effacement. *A* Site of resistant lesion, *B* angioplasty balloon guidewire, *C* second guidewire being used as cutting wire

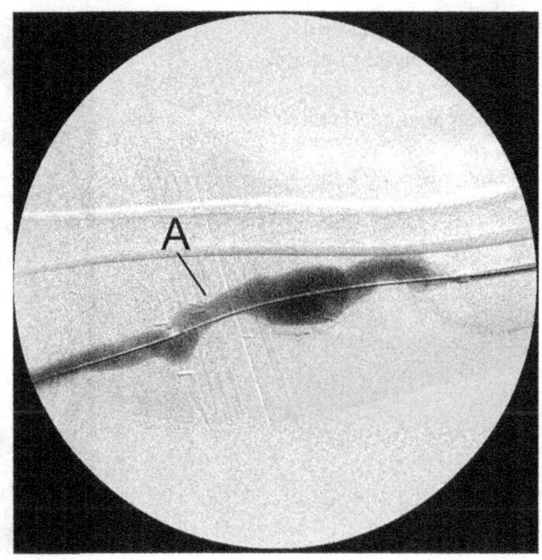

Fig. 42.6 Post-procedure angiogram. *A* Site of previous lesion

In some cases, multiple prolonged dilatations can weaken a resistant lesion to the point that it eventually gives way. However, there are techniques available that are specifically designed to weaken the integrity of a stenotic lesion. These involve cutting it in some manner. The use of a cutting balloon has been advocated by some interventionalists for this problem [2–6]; however, there are techniques that may be effective and are much more economical, one of these is the parallel guidewire technique that was used in this case.

The concept of this technique is to use a guidewire as a cutting device to weaken the resistant lesion. It has been referred to using several different terms – "buddy-wire angioplasty [7]," "parallel wire angioplasty [8]," and "focused force angioplasty [9–11]." Most of the literature dealing with this technique relates to coronary artery angioplasty; however, it has also been used to treat resistant lesions associated with dialysis vascular access [12, 13]. There are two ways that this can be accomplished – by using a single guidewire that has been buckled back on itself to create a parallel segment [13] or by using a second guidewire as was done in this case. In doing this, the author has found that a stainless steel guidewire seems to work better than one of the hydrophilic coated guidewires.

With the parallel guidewire technique, a second guidewire is passed (Fig. 42.4) so that it lies parallel to the angioplasty balloon across the resistant lesion. The balloon is then dilated to maximal pressure pressing the second guidewire into the wall of the lesion to disrupt it (Fig. 42.5). In one series of 22 cases [13], this technique was successful in all instances. In 18 of the cases, satisfactory results were obtained with a pressure lower than what had been used for the initial unsuccessful angioplasty. In the remaining four cases, the technique was implemented several times using maximum dilatation pressures, changing the guidewire position before each additional inflation.

References

1. Trerotola SO, Stavropoulos SW, Shlansky-Goldberg R, et al. Hemodialysis-related venous stenosis: treatment with ultrahigh-pressure angioplasty balloons. Radiology. 2004;231:259.
2. Bittl JA, Feldman RL. Cutting balloon angioplasty for undilatable venous stenoses causing dialysis graft failure. Catheter Cardiovasc Interv. 2003;58:524.
3. Song HH, Kim KT, Chung SK, et al. Cutting balloon angioplasty for resistant venous stenoses of Brescia-Cimino fistulas. J Vasc Interv Radiol. 2004;15:1463.
4. Singer-Jordan J, Papura S. Cutting balloon angioplasty for primary treatment of hemodialysis fistula venous stenoses: preliminary results. J Vasc Interv Radiol. 2005;16:25.
5. Kariya S, Tanigawa N, Kojima H, et al. Percutaneous transluminal cutting-balloon angioplasty for hemodialysis access stenoses resistant to conventional balloon angioplasty. Acta Radiol. 2006;47:1017.
6. Kariya S, Tanigawa N, Kojima H, et al. Primary patency with cutting and conventional balloon angioplasty for different types of hemodialysis access stenosis. Radiology. 2007;243:578.
7. Selig MB. Lesion protection during fixed-wire balloon angioplasty: use of the "buddy wire" technique and access catheters. Catheter Cardiovasc Diagn. 1992;25:331.
8. Yazdanfar S, Ledley GS, Alfieri A, et al. Parallel angioplasty dilatation catheter and guide wire: a new technique for the dilatation of calcified coronary arteries. Catheter Cardiovasc Diagn. 1993;28:72.
9. Stillabower ME. Longitudinal force focused coronary angioplasty: a technique for resistant lesions. Catheter Cardiovasc Diagn. 1994;32:196.
10. Solar RJ, Ischinger TA. Focused force angioplasty: theory and application. Cardiovasc Radiat Med. 2003;4:47.
11. Meerkin D, Lee SH, Tio FO, et al. Effects of focused force angioplasty: pre-clinical experience and clinical confirmation. J Invasive Cardiol. 2005;17:203.
12. Pua U. Focused force fistuloplasty as an adjunct to conventional balloon fistuloplasty in resistant stenosis. Nephrology (Carlton). 2012;17:97.
13. Fukasawa M, Matsushita K, Araki I, et al. Self-reversed parallel wire balloon technique for dilating unyielding strictures in native dialysis fistulas. J Vasc Interv Radiol. 2002;13:943.

Chapter 43
Cephalic Vein Outflow Relocation

Gerald A. Beathard

Case Presentation

The patient was a 58-year-old male who had been on dialysis for 4 years with a brachiocephalic fistula. He has had recurrent cephalic arch stenosis for the past 18 months and has been treated with angioplasty on the average of every 2 months. He was referred to the vascular access center because of decreased blood flow and a hyperpulsatile fistula. Upon examination the patient was found to have a brachiocephalic fistula in his left upper arm, which was dilated and hyperpulsatile with a thrill palpable in the right subclavicular area.

An angiogram was performed which showed that the fistula was dilated and had a relatively broad zone of irregular stenosis at the cephalic arch (Fig. 43.1). Because of the frequency with which angioplasty was required, a decision was made to refer the patient to surgery with the recommendation of doing a cephalic vein outflow relocation procedure.

Six months later, the patient was again referred to the vascular access center because of his complaint of hand pain. Examination of the fistula revealed no abnormalities (Fig. 43.2). An angiogram performed at that time revealed the fistula was dilated and the surgical revision site looked good (Fig. 43.3). Further evaluation of the patient's hand pain revealed that it was secondary to carpal tunnel syndrome.

G.A. Beathard, MD, PhD (✉)
Lifetime Vascular Access, University of Texas Medical Branch,
5135 Holly Terrace, Houston, TX 77056, USA
e-mail: gbeathard@msn.com

© Springer International Publishing AG 2017
A.S. Yevzlin et al. (eds.), *Dialysis Access Cases*,
DOI 10.1007/978-3-319-57500-1_43

Fig. 43.1 Angiogram of
cephalic arch showing
zone of stenosis. *A* Zone of
stenosis

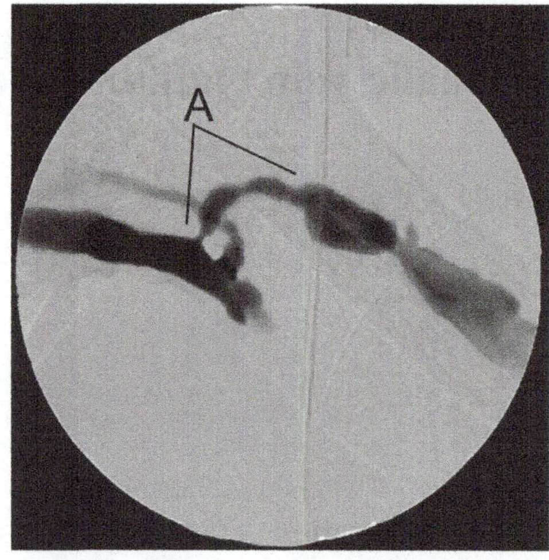

Fig. 43.2 Appearance of
brachiocephalic fistula 6
months after cephalic
outflow relocation surgery

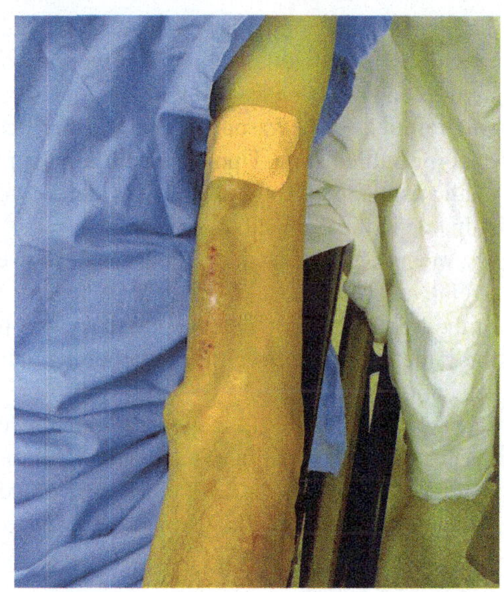

Discussion

The cephalic arch is a rather unique vascular structure for three reasons. First, the anatomical configuration of the cephalic arch makes it one of the unique sites in association with hemodialysis access that has been referred to as a "swing point" [1]. Flow at these points is non-laminar and is associated with decreased shear stress [2]. This results in a cascade of events that result in neointimal hyperplasia causing venous stenosis [3, 4]. Second, one or more valves are located in the cephalic arch. It is well known that venous stenosis has a predilection for sites of valves [5], and

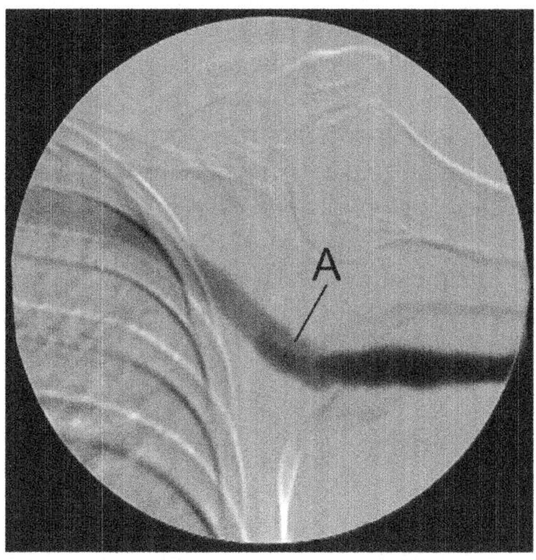

Fig. 43.3 Angiogram of brachiocephalic fistula 6 months after cephalic outflow relocation surgery. *A* Site of surgical anastomosis

while the location of valves within veins is somewhat variable, their location in this segment of the cephalic vein is rather constant. Third, the cephalic arch is mechanically limited in its ability to dilate in response to an increase in blood flow commensurate with the formation of a functioning arteriovenous fistula (AVF) because of anatomical relationships associated with the clavipectoral fascia. This is a dense membrane occupying the interval between the pectoralis minor muscle and the subclavian vessels. It is pierced by the cephalic vein, thoracoacromial artery and vein, and lateral pectoral nerve. The density of this fascial membrane and the fact that it is accompanied by other structures of significant size as it passes though the foramen may prove problematic when this vein is called upon to dilate.

There are three alternatives to treatment – angioplasty, angioplasty with stenting, and surgery. When treated with angioplasty, lesions in the cephalic vein have been found to be resistant to treatment and recurrences are frequent [6]. Management with angioplasty only is reasonable if the frequency of recurrence is not too great. Stenting may also be beneficial, although it should not be regarded as a permanent treatment [7]. The standard surgical approach has been to expose the cephalic arch and apply patch angioplasty at the site of the lesion. This has the drawback of being a major surgical procedure. A novel surgical approach was reported by Chen et al. [8] which they described as a venovenostomy but has come to be referred to by many as "outflow relocation." According to this technique, the cephalic vein is dissected as far proximal as possible to ensure adequate length. It is then transposed through a subcutaneous tunnel and anastomosed to the upper basilic/axillary vein in an end-to-side manner. In their report, seven patients with brachiocephalic fistula and cephalic arch stenosis were transposed. A primary patency of 70% was observed at 6 months and 60% at 12 months. It should be mentioned that this venovenostomy anastomotic site can develop stenosis requiring treatment, but this responds better than the cephalic arch lesion [9].

A more detailed study of the outflow relocation procedure was reported [9]. This report evaluated the frequency of angioplasty of the cephalic arch lesion before relocation surgery with that of the venovenostomy anastomosis after surgery in 13 cases. Primary patency rates for angioplasty before surgical revision were 23%, 8%, and 0% at 3, 6, and 12 months, respectively. Following surgical revision, all patients needed angioplasty procedures. However, primary patency increased to 92%, 69%, and 39% at 3, 6, and 12 months, respectively ($p = 0.0001$). Secondary patency before the surgical revision at 3, 6, and 12 months was 100%, 39%, and 8%, respectively, compared with 92% at 3, 6, and 12 months postsurgical revision ($p = 0.0003$).

References

1. Falk A, Teodorescu V, Lou WY, Uribarri J, Vassalotti JA. Treatment of "swing point stenoses" in hemodialysis arteriovenous fistulae. Clin Nephrol. 2003;60(1):35–41.
2. Ene-Iordache B, Remuzzi A. Disturbed flow in radial-cephalic arteriovenous fistulae for haemodialysis: low and oscillating shear stress locates the sites of stenosis. Nephrol Dial Transplant. 2011;27(1):358–68.
3. Asif A, Roy-Chaudhury P, Beathard GA. Early arteriovenous fistula failure: a logical proposal for when and how to intervene. Clin J Am Soc Nephrol. 2006;1(2):332–9.
4. Badero OJ, Salifu MO, Wasse H, Work J. Frequency of swing-segment stenosis in referred dialysis patients with angiographically documented lesions. Am J Kidney Dis. 2008;51(1):93–8.
5. Glanz S, Bashist B, Gordon DH, Butt K, Adamsons R. Angiography of upper extremity access fistulas for dialysis. Radiology. 1982;143(1):45–52.
6. Rajan DK, Clark TW, Patel NK, Stavropoulos SW, Simons ME. Prevalence and treatment of cephalic arch stenosis in dysfunctional autogenous hemodialysis fistulas. J Vasc Interv Radiol. 2003;14(5):567–73.
7. Shemesh D, Goldin I, Zaghal I, Berlowitz D, Raveh D, Olsha O. Angioplasty with stent graft versus bare stent for recurrent cephalic arch stenosis in autogenous arteriovenous access for hemodialysis: a prospective randomized clinical trial. J Vasc Surg. 2008;48(6):1524–31, 1531e1–2.
8. Chen JC, Kamal DM, Jastrzebski J, Taylor DC. Venovenostomy for outflow venous obstruction in patients with upper extremity autogenous hemodialysis arteriovenous access. Ann Vasc Surg. 2005;19(5):629–35.
9. Kian K, Unger SW, Mishler R, Schon D, Lenz O, Asif A. Role of surgical intervention for cephalic arch stenosis in the "fistula first" era. Semin Dial. 2008;21(1):93–6.

Chapter 44
Venous Hypertension of Hand Associated with Arteriovenous Fistula

Gerald A. Beathard

Case Presentation

The patient was a 53-year-old male hemodialysis patient with a radial-cephalic fistula (AVF) in his left forearm. The AVF had been in place for 3 years and has been functioning well. The patient was referred to the dialysis access center because of complaints of edema and paresthesia of his ipsilateral hand.

Physical examination revealed that the patient had multiple large and large, dilated, pulsatile veins on the back of his hand (Fig. 44.1). A palpable thrill was present in the veins distal to the fistula. By manually occluding the major vein proximally and distally in a sequential fashion, it was demonstrated that the blood flow in the vein was retrograde. When the AVF anastomosis was occluded, the veins on the back of the hand collapsed.

After the patient was prepped and draped, the AVF was cannulated approximately 15 cm above the anastomosis in a retrograde direction. An antegrade angiogram was performed through a catheter, which was positioned with the tip in the lower portion of the AVF. This showed a large vein originating approximately 4 cm above anastomosis, which extended down into the hand with a number of branches (Fig. 44.2, arrow indicates anastomosis). After this vein was surgically ligated, the AVF was manually occluded in order to obtain retrograde flow and a second angiogram was performed. This showed another large vein originating at the anastomosis extending down into the hand (Fig. 44.3, arrow indicates anastomosis). This vein was also surgically ligated. At this point the veins on the back of the patient's hand were collapsed. A follow-up angiogram was performed which showed good flow in the fistula with no blood flow into the ligated veins (Fig. 44.4, arrow indicates anastomosis).

G.A. Beathard, MD, PhD (✉)
Lifetime Vascular Access, University of Texas Medical Branch,
5135 Holly Terrace, Houston, TX 77056, USA
e-mail: gbeathard@msn.com

© Springer International Publishing AG 2017
A.S. Yevzlin et al. (eds.), *Dialysis Access Cases*,
DOI 10.1007/978-3-319-57500-1_44

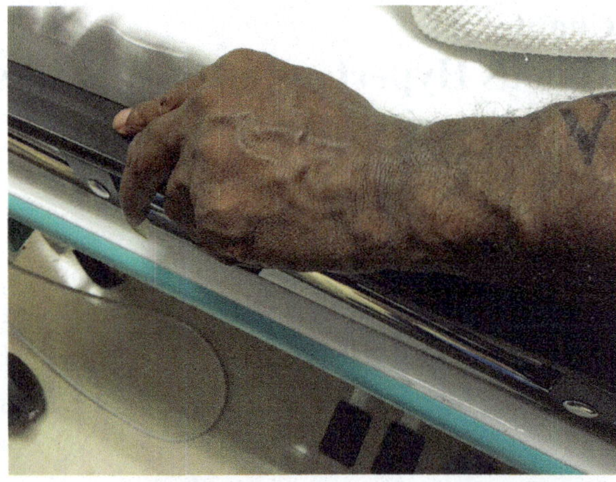

Fig. 44.1 Appearance of patient's hand showing large distended veins

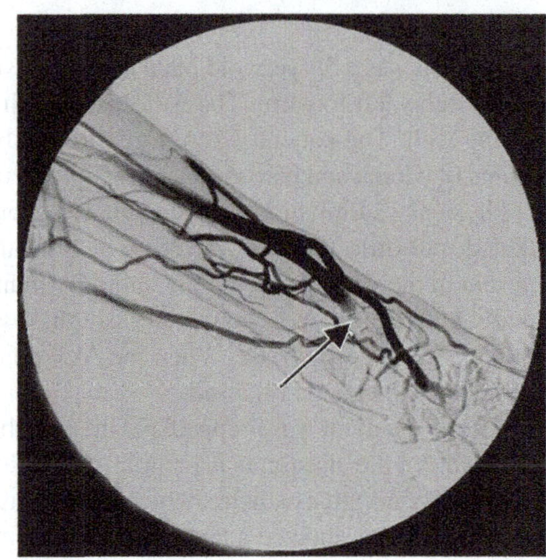

Fig. 44.2 Angiogram showing the vein arising approximately 4 cm above anastomosis and extending down into the hand (*arrow* indicates anastomosis)

Discussion

Symptomatic peripheral venous hypertension of the hand secondary to retrograde outflow from an AVF is a rare condition. It has generally been reported in association with a radial-cephalic AVF created by side-to-side anastomosis [1–4]. This was the technique originally used to create the Cimino fistula [5]; however, even with that type of AVF, it was estimated to occur in less than 1% of cases [6, 7]. In a study involving snuffbox radial-cephalic AVFs, 60 patients were divided into two groups of equal size [8]. In one group, the distal vein was ligated at the time of surgery. In the other, the distal vein was left patent. Venous hypertension in the hand developed in two (6.6%) cases in the group in which the vein was not ligated. None was seen in the ligated group.

Fig. 44.3 Angiogram performed with manual occlusion of fistula to create retrograde flow after ligation of the first vein (*arrow* indicates anastomosis)

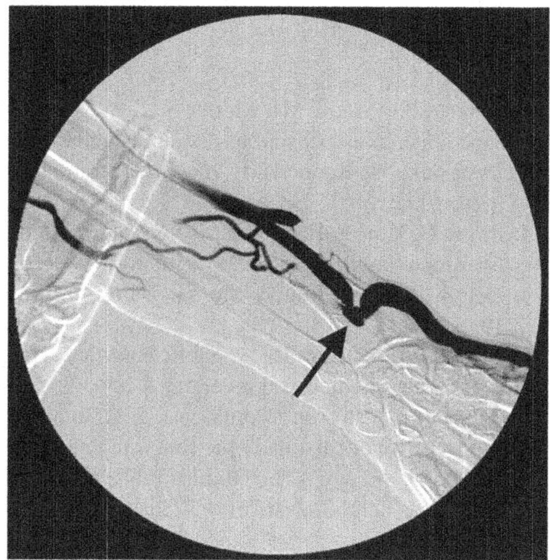

Fig. 44.4 Angiogram of fistula following ligation of vein distal to the anastomosis (*arrow* indicates anastomosis)

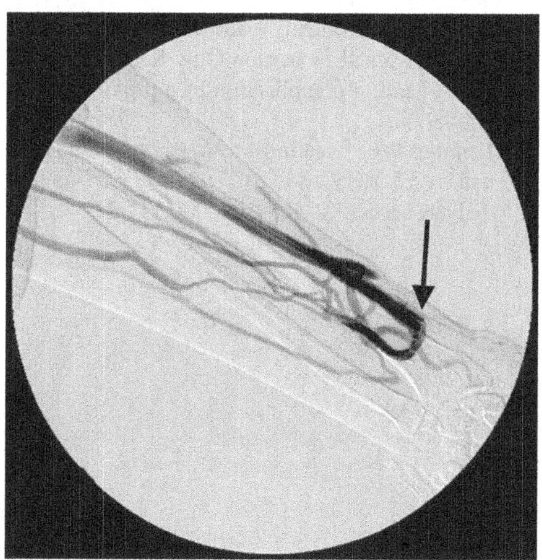

The clinical features exhibited vary from venous distention with mild swelling of the hand to venous varicosities and in severe cases, ulceration. In an early publication, a syndrome referred to as "sore thumb syndrome" was described, characterized by a swollen, cyanotic thumb with eczematous skin changes resulting from AVF-associated venous hypertension of the hand [9].

Most of the cases reported in the more recent literature have been in association with bidirectional AVFs created using proximal radial inflow [10] and then generally only with downstream stenosis [11].

The treatment for this problem is ligation of the veins distal to the anastomosis [1–4] as was done in this case. This results in resolution of the venous hypertension; problems such as ulceration that may be associated with severe cases will generally resolve following this treatment. In order to avoid this complication, an end-to-side anastomosis is recommended [8].

References

1. Kootstra G, Slooff MJ, Meijer S, Tegzess AM. Venous hypertension of the hand caused by subcutaneous arteriovenous fistulae established for hemodialysis. Arch Chir Neerl. 1979;31:43.
2. Deshmukh N, Reppert M. Venous ulceration of the hand secondary to a cimino fistula. Mil Med. 1993;158:752.
3. Irvine C, Holt P. Hand venous hypertension complicating arterio-venous fistula construction for haemodialysis. Clin Exp Dermatol. 1989;14:289.
4. Delpin EA. Swelling of the hand after arteriovenous fistula for hemodialysis. Am J Surg. 1976;132:373.
5. Brescia MJ, Cimino JE, Appel K, Hurwich BJ. Chronic hemodialysis using venipuncture and a surgically created arteriovenous fistula. N Engl J Med. 1966;275:1089.
6. Kinnaert P, Struyven J, Mathieu J, et al. Intermittent claudication of the hand after creation of an arteriovenous fistula in the forearm. Am J Surg. 1980;139:838.
7. Adams E, Sidawy A. In: Rutherford R, editor. Nonthrombotic complications of arteriovenous access for hemodialysis in: vascular surgery. 6th ed. Philadelphia: Elsevier Saunders; 2005. p. 1692.
8. Beigi AA, Masoudpour H, Alavi M. The effect of ligation of the distal vein in snuff-box arteriovenous fistula. Saudi J Kidney Dis Transpl. 2009;20:1110.
9. Butt KM, Friedman EA, Kountz SL. Angioaccess. Curr Probl Surg. 1976;13:1.
10. Anaya-Ayala JE, Charlton-Ouw KM, Cardon AL, Peden EK. Peripheral venous hypertension of the hand: a complication of a proximal radial artery arteriovenous fistula. J Vasc Access. 2008;9:64.
11. Jennings WC. Creating arteriovenous fistulas in 132 consecutive patients: exploiting the proximal radial artery arteriovenous fistula: reliable, safe, and simple forearm and upper arm hemodialysis access. Arch Surg. 2006;141:27.

Chapter 45
Grade I Extravasation

Gerald A. Beathard

Case Presentation

The patient was a 61-year-old female dialysis patient with a left forearm loop arteriovenous graft (AVG). The patient was referred to the vascular access center because of poor blood flow in the access. Physical examination revealed a loop AVG in the left forearm, which was hyper-pulsatile. The thrill over the venous anastomosis was systolic only. The bruit was high pitched and systolic only. The drainage was primarily through the basilic vein, which was hyper-pulsatile, and a localized thrill was palpable in the general region of the axilla.

After the patient was prepped and draped, an angiogram was performed which showed marked stenosis at the venous anastomosis and a second area of stenosis in the axillary vein. After the guidewire was passed, an attempt was made to treat the lesion in the axillary vein using a high-pressure 9 × 4 angioplasty balloon. Full effacement of the balloon could not be achieved; a narrow circumferential defect in the balloon could not be eliminated (Fig. 45.1, arrow). An ultrahigh-pressure balloon was inserted, and the lesion was dilated using 30 atmospheres pressure. Venous rupture occurred resulted in extravasation of radiocontrast (Fig. 45.2, arrows). Blood flow was immediately obstructed manually at the level of the venous anastomosis. A hematoma could be felt in the area of the lesion. Radio contrast was injected, and unobstructed blood flow through the area of extravasation was noted. The extravasation was stable with no continuing leakage of radiocontrast (Fig. 45.3, arrows indicate area of extravasation). Over 5 min of observation, no further extravasation was noted. The venous anastomosis lesion was then treated using an 8 × 4 angioplasty balloon with good result. The area of extravasation was again examined and found to have not changed. The patient was returned to the dialysis facility for dialysis.

G.A. Beathard, MD, PhD (✉)
Lifetime Vascular Access, University of Texas Medical Branch,
5135 Holly Terrace, Houston, TX 77056, USA
e-mail: gbeathard@msn.com

© Springer International Publishing AG 2017
A.S. Yevzlin et al. (eds.), *Dialysis Access Cases*,
DOI 10.1007/978-3-319-57500-1_45

Fig. 45.1 Appearance of
angioplasty balloon with
resistant lesion preventing
total effacement (*arrow*)

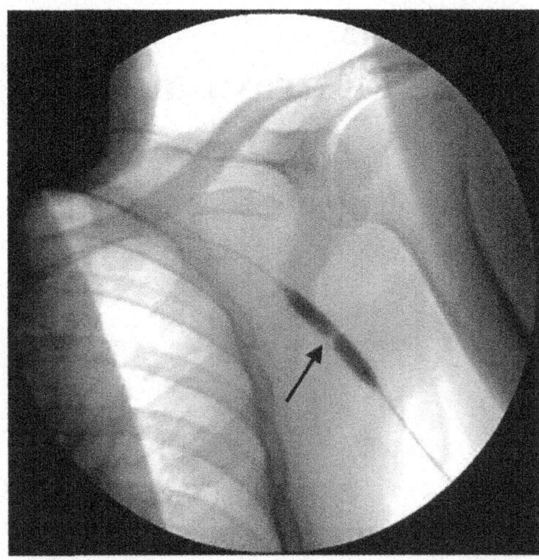

Fig. 45.2 Angiogram
showing extravasation of
radiocontrast (*arrows*)

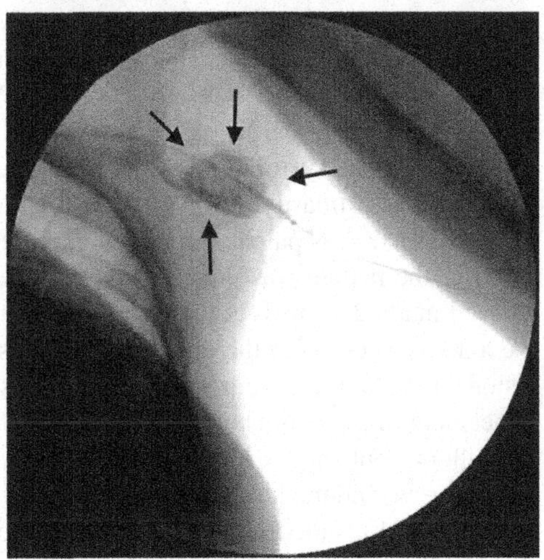

Discussion

The most frequent procedure-related complication seen in association with angioplasty that dictates the need for intervention is vein rupture with extravasation [1];
however, this is not a frequent occurrence. In a series of 5121 angioplasty procedures performed on dialysis access, an overall complication rate of 3.54% was seen.
This series was composed of AVGs, 3560 cases with a 1.15% complication rate, and
arteriovenous fistula (AVF), 1561 cases with a 4.48% complication rate. Seventy
percent of these complications were vein rupture with extravasation of some degree
[1]. In two other large studies, each one of which involved over 1200 cases, a 0.9%

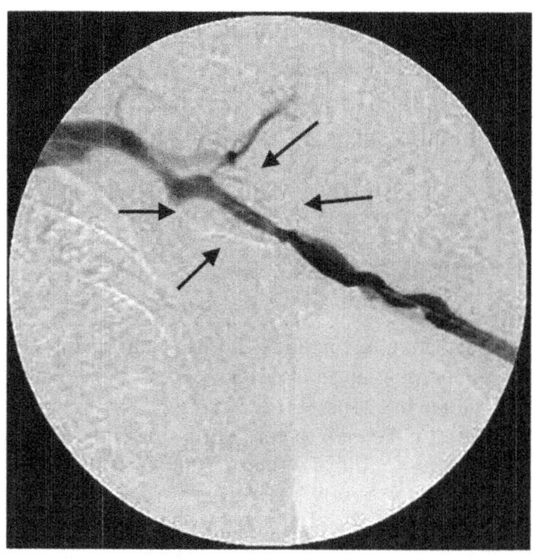

Fig. 45.3 Angiogram showing good blood flow through area of extravasation (indicated by *arrows*) with no continuing extravasation of radiocontrast

incidence [2] and a 2% incidence [3], of vein rupture with extravasation was reported, respectively.

The clinical significance of vein rupture with extravasation is variable, ranging from none to disaster for the access. The difference lies in the severity of the tear. The presence of this complicating event is generally heralded by the extravasation of radiocontrast, blood, or both. Vein rupture with extravasation becomes more obvious when there is the formation of an obvious hematoma. The amount of extravasated contrast associated with the hematoma may be minimal or absent. In actuality the size of the hematoma is not particularly important in determining the sequelae with which it is associated. One may see a small degree of extravasation with hematoma formation as a result of complete vein disruption that was quickly controlled. On the other hand, a very large hematoma may be completely stable having no effect on blood flow. It is useful to use a classification system for extravasation based upon its clinical significance [4–6].

When an extravasation is noted, one must make two determinations – (1) is it stable or continuing to enlarge and (2) does it affect flow? As soon as it is obvious that a vascular rupture has occurred and a hematoma is forming, the access should be manually occluded upstream (distal) to the site of rupture to relieve pressure on the site and limit the degree of extravasation until a determination of its severity can be determined.

This most common type of extravasation is a Grade I extravasation and is by definition stable, e.g., not continuing to grow and does not affect flow. Although the Grade I situation may cause concern on the part of the operator and the patient, it is of no real consequence to the outcome of the procedure and requires no specific treatment. This is true regardless of its size. In general, a hematoma that remains stable over 30 min to an hour period will continue to behave in this manner as long as there is no downstream obstruction. It does not require further observation.

The patient with a Grade I extravasation will probably have an ecchymosis which may be quite large, depending on the size of the extravasation. Localized discomfort may be significant and may last for several days. They may require symptomatic treatment measures [5]. Mild analgesics and a heating pad may be helpful.

References

1. Beathard GA, Litchfield T, Physician Operators Forum of RMS Lifeline Inc. Effectiveness and safety of dialysis vascular access procedures performed by interventional nephrologists. Kidney Int. 2004;66:1622.
2. Bittl JA. Venous rupture during percutaneous treatment of hemodialysis fistulas and grafts. Catheter Cardiovasc Interv. 2009;74:1097.
3. Pappas JN, Vesely TM. Vascular rupture during angioplasty of hemodialysis graft-related stenoses. J Vasc Access. 2002;3:120.
4. Beathard GA. Angioplasty for arteriovenous grafts and fistulae. Semin Nephrol. 2002;22:202.
5. Beathard GA. Management of complications of endovascular dialysis access procedures. Semin Dial. 2003;16:309.
6. Beathard GA, Urbanes A, Litchfield T. The classification of procedure-related complications–a fresh approach. Semin Dial. 2006;19:527.

Part V
Central Veins

Chapter 46
Compensated Central Vein Occlusion

Gerald A. Beathard

Case Presentation

The patient was a 68-year-old male who had been on hemodialysis for several years. During this period of time, he had had a number of central venous catheters which had been in place for prolonged periods. He also received several upper extremity arteriovenous accesses which had failed. Approximately 4 months earlier, a successful brachial-basilic fistula was created in his left arm and had been used effectively for dialysis treatments. After the fistula was placed, the patient developed marked swelling of his arm. Over the next few months, however, the swelling had become less and recently has completely disappeared without any intervention. The patient has been dialyzing well and has a sp. Kt/V of 1.7. The patient was referred to the vascular access center because of the presence of numerous large collateral veins over his upper arm and chest.

Examination of the patient revealed that he had a functioning left brachial-basilic fistula which was dilated but not hyper-pulsatile. Numerous large superficial tortuous veins were present over the patient's upper arm, neck, chest, and abdomen (Figs. 46.1 and 46.2). After the patient was prepped and draped, the fistula was cannulated, and an angiogram was performed. This showed that the basilic vein was patent and markedly dilated with many collaterals. The left brachiocephalic vein was patent up to the expected point of junction with the superior vena cava where there was a total occlusion (Figs. 46.3 and 46.4). Additionally, the jugular venous system in the neck was markedly dilated and tortuous as was the internal thoracic vein on the right (Figs. 46.5 and 46.6). At this point the procedure was discontinued.

G.A. Beathard, MD, PhD (✉)
Lifetime Vascular Access, University of Texas Medical Branch,
5135 Holly Terrace, Houston, TX 77056, USA
e-mail: gbeathard@msn.com

© Springer International Publishing AG 2017
A.S. Yevzlin et al. (eds.), *Dialysis Access Cases*,
DOI 10.1007/978-3-319-57500-1_46

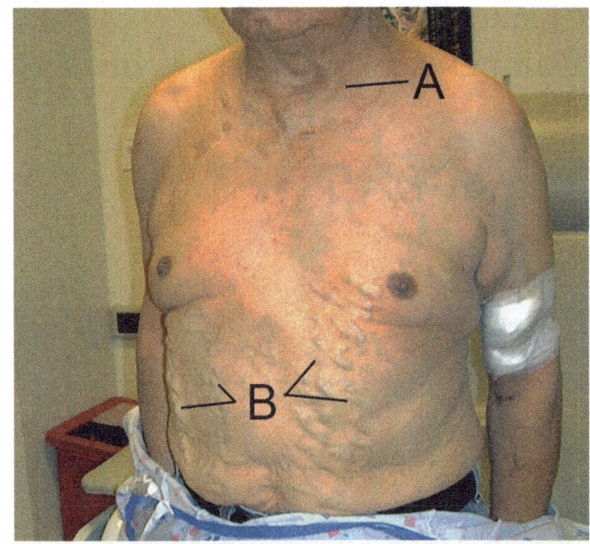

Fig. 46.1 Collateral veins on surface of patients body. *A* dilated jugular venous system, *B* dilated veins over patient's abdomen

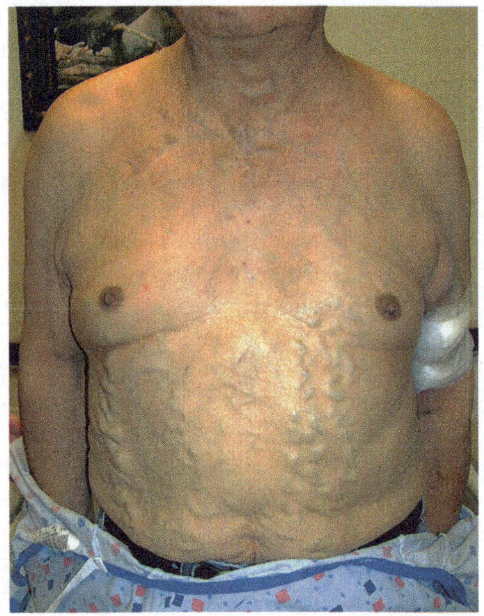

Fig. 46.2 Collateral veins on surface of patients body, second view

Discussion

As is frequently the case, this patient has developed central vein stenosis from the chronic use of dialysis catheters. This has led to total obstruction which on the side of his current arteriovenous access is at the level of the junction between the bra-chiocephalic vein and the superior vena cava. Unfortunately, early signs of the

Fig. 46.3 Angiogram of
brachial-basilic fistula.
A fistula

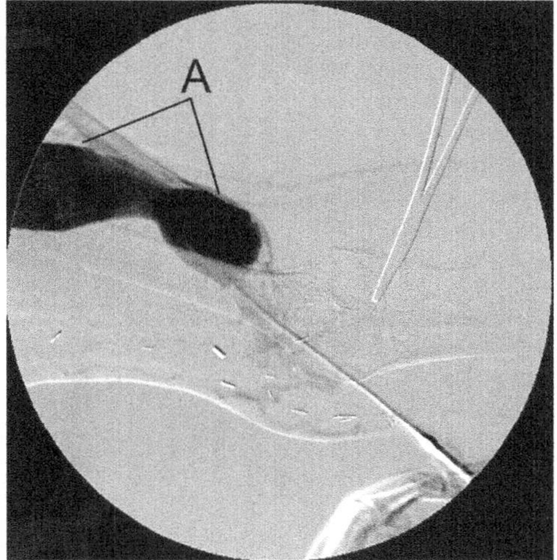

Fig. 46.4 Angiogram of
upper fistula and central
veins. *A* internal jugular
vein, *B* subclavian vein;
C, basilic vein/axillary
vein with collaterals

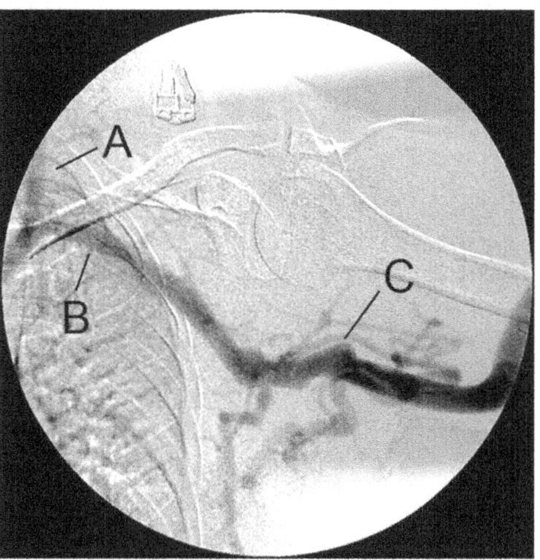

problem associated with the progressive edema of the ipsilateral extremity were
ignored. With the development of collaterals secondary to the venous obstruction,
the patient eventually decompressed his central venous circulation. By physical
examination, pressure within his access appeared to be relatively normal, and his
dialysis has been proceeding without complication.

While the development of numerous collaterals secondary to venous obstruction
is to be expected, decompression of the obstruction to the degree apparent in this
patient is not common nor is it rare. Although the patient is dialyzing without complication at this time and is very likely to continue to do so, he is at considerable risk
due to the size and number of superficial venous collaterals that are present. Should
the patient be involved in an accident or require surgery, these vessels could lead to
serious, life-threatening bleeding.

  G.A. Beathard

Fig. 46.5 Angiogram of central veins showing collaterals, another view. *A* point of obstruction, *B* internal jugular vein, *C* anterior jugular vein and jugular arch, *D* subclavian vein, *E* brachiocephalic vein

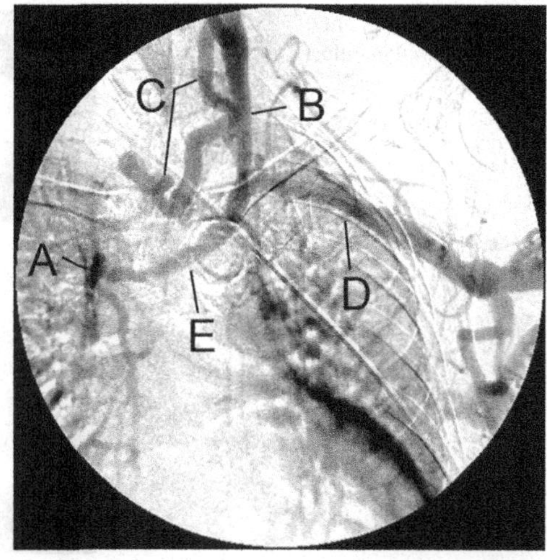

Fig. 46.6 Angiogram of central veins showing collaterals, another view. *A* internal thoracic vein

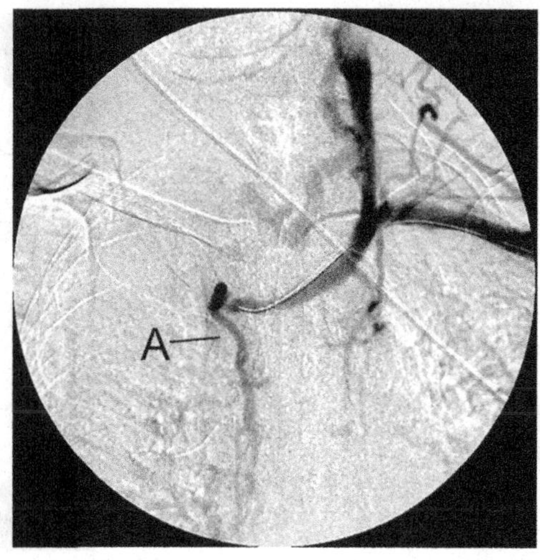

Chapter 47
Central Vein Stenosis Accessed from Femoral Vein

Gerald A. Beathard

Case Presentation

The patient was a 56-year-old female with the right upper arm arteriovenous graft (AVG). She was known to have brachiocephalic stenosis from an angiogram that had been done 5 months previously. The lesion was not treated at that time because of inability to cross the lesion with a guidewire. The patient was referred to the vascular access center at this time because of marked right arm, face, and breast swelling that had been progressing over the past week.

Examination of the patient revealed that she had a right upper arm AVG which was hyper-pulsatile. There was marked swelling of the arm making palpation of the access difficult. There was evidence of collateral veins over the anterior chest.

After the patient was prepped and draped, the AVG was cannulated, and an angiogram was performed. This showed what appeared to be complete obstruction at the junction of the brachiocephalic and subclavian vein (Fig. 47.1). Multiple attempts were made to cross this lesion with a guidewire including using a vascular catheter to assist the process. All attempts were unsuccessful. The femoral vein was then cannulated, and a catheter was advanced up to the site of the lesion. Using a hydrophilic guidewire assisted with the use of a vascular catheter, the lesion was crossed, and an angiogram was performed to confirm the position (Fig. 47.2). The guidewire was further advanced down to the level of the venous anastomosis of the AVG. The distal end of the guidewire was captured with a snare and brought out through the sheath that had been previously placed in the AVG (Fig. 47.3). The stenotic lesion was then dilated using a 14 × 4 angioplasty balloon (Figs. 47.4 and 47.5). A post-procedure angiogram was performed which showed good result from the treatment (Fig. 47.6).

G.A. Beathard, MD, PhD (✉)
Lifetime Vascular Access, University of Texas Medical Branch,
5135 Holly Terrace, Houston, TX 77056, USA
e-mail: gbeathard@msn.com

© Springer International Publishing AG 2017
A.S. Yevzlin et al. (eds.), *Dialysis Access Cases*,
DOI 10.1007/978-3-319-57500-1_47

Fig. 47.1 Initial angiogram showing side of stenosis. *A* site of stenosis

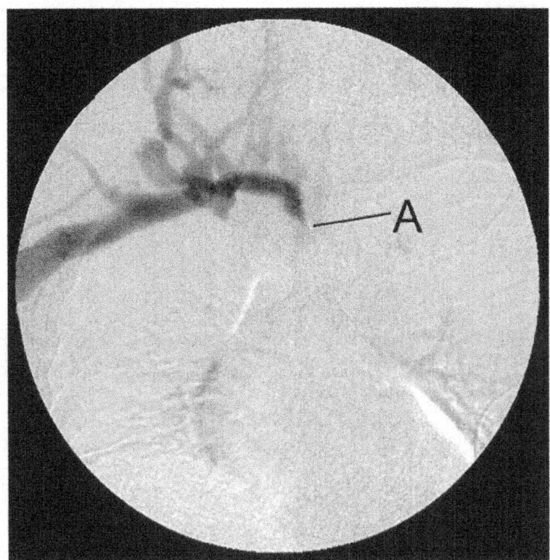

Fig. 47.2 Crossing stenotic site with vascular catheter guidewire advanced from femoral cannulation. *A* - subclavian vein peripheral to stenosis, *B* - guidewire coming from femoral insertion site

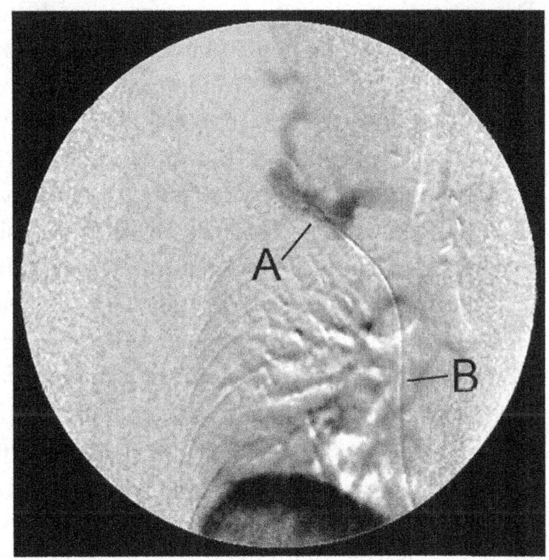

The patient was returned to the dialysis facility where she was dialyzed with difficulty. The swelling of her arm resolved over the next 4 days.

Discussion

This case demonstrates the fact that, in an instance in which it is not possible to advance a guidewire across a severely stenotic lesion that appears to be a total occlusion, one should not conclude that it is untreatable until an attempt is made from the opposite direction. By cannulating the femoral vein and passing a vascular catheter up to the stenotic site, it was possible to cross the lesion and successfully treat it.

Fig. 47.3 Snaring
guidewire advanced from
femoral site. *A* snare and
tip of guidewire

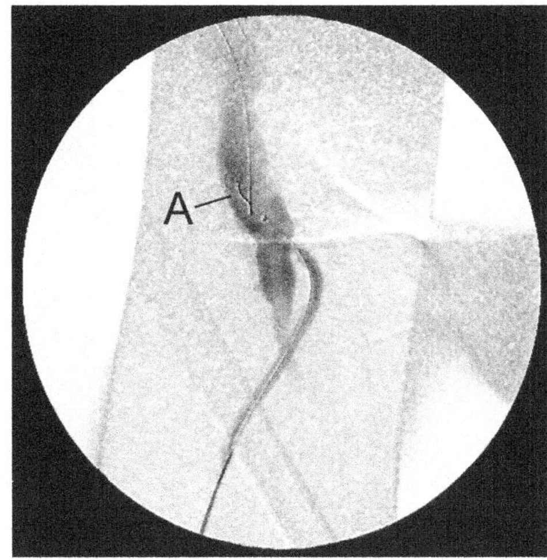

Fig. 47.4 Angioplasty of
stenotic lesion

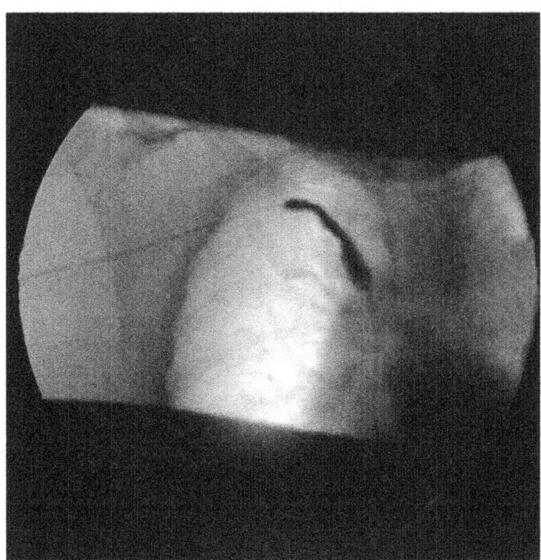

It is probable that the angioplasty could've been done from the femoral cannulation site. However, bringing the wire out the cannulation site in the access does make the performance of angioplasty more easily accomplished.

Lesions of this type do not occur quickly, the mere fact that it was so advanced suggests that this patient was not being monitored for vascular access problems very effectively. Addressing central venous lesions on a more timely basis can be expected to prevent such occurrences.

Fig. 47.5 Angioplasty of
stenotic lesion showing
total effacement of the loan

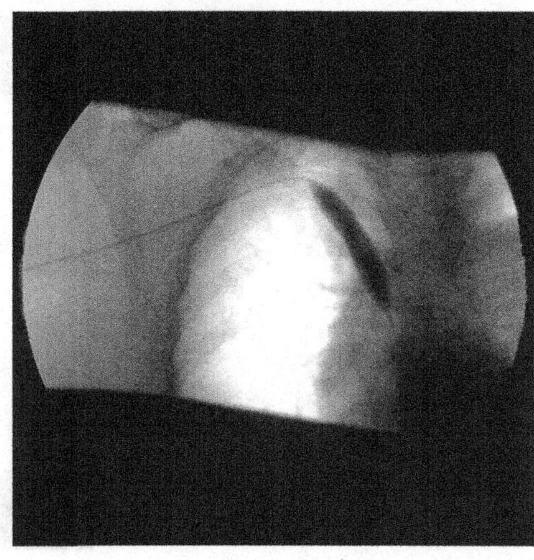

Fig. 47.6 Post-procedure
angiogram showing good
result from treatment.
A site of stenosis

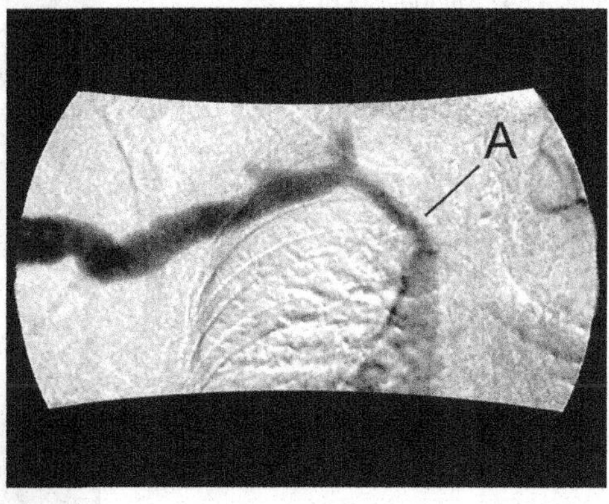

Chapter 48
Patient with Intractable Pain in Arteriovenous Graft

Gerald A. Beathard

Case Presentation

The patient was a 52-year-old female who had been receiving hemodialysis treatments for 18 months using a right forearm arteriovenous graft (AVG). The patient had carpal tunnel syndrome in her right hand of moderate severity. For 4 months she had been complaining of severe pain in the area of her graft, which has gotten progressively worse. The pain was constant, on the medial aspect (venous side of graft) side of the forearm. The pain was clearly distinct from carpal tunnel pain. Because of the severity of the pain and her inability to achieve any relief, she has been referred to pain management.

The patient was referred to the dialysis access center because of poor access of blood flow on dialysis. Physical examination reveals that she has an increased pulse in the graft with a localized thrill at the venous anastomosis. The bruit in this area is high pitched and systolic only. There was a large dilated vein arising in the approximate area of the venous anastomosis, which was pulsatile.

The patient was prepped and draped. The AVG was cannulated in an antegrade direction low on the venous side. An angiogram was performed which showed a tight stenosis just above the anastomosis (Figs. 48.1 and 48.2). There were a number of large collateral veins in the area which seemed to originate from a large vein that arose between the venous end of the AVG and the point of stenosis. The lesion was treated with an 8 × 4 angioplasty balloon with good result.

Post-angioplasty, the patient reported that her forearm pain was completely gone. Four months later, the patient reported that although mild, the pain had recurred. She was reevaluated, and it was found that the stenotic lesion had recurred along

G.A. Beathard, MD, PhD (✉)
Lifetime Vascular Access, University of Texas Medical Branch,
5135 Holly Terrace, Houston, TX 77056, USA
e-mail: gbeathard@msn.com

© Springer International Publishing AG 2017
A.S. Yevzlin et al. (eds.), *Dialysis Access Cases*,
DOI 10.1007/978-3-319-57500-1_48

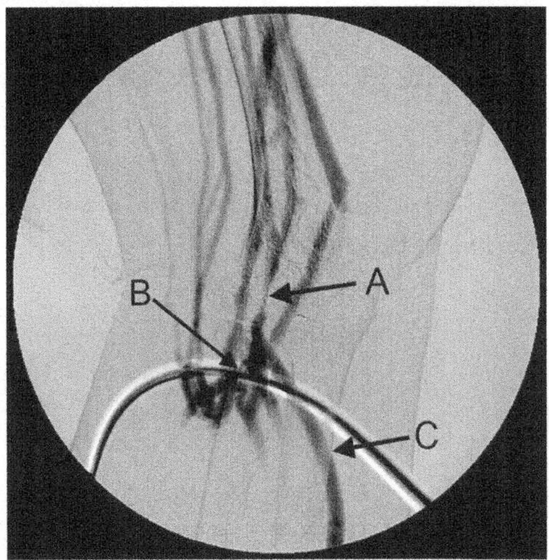

Fig. 48.1 Angiogram showing stenotic lesion (*A*), collateral originating between venous end of graft and stenosis (*B*), and venous end of graft (*C*)

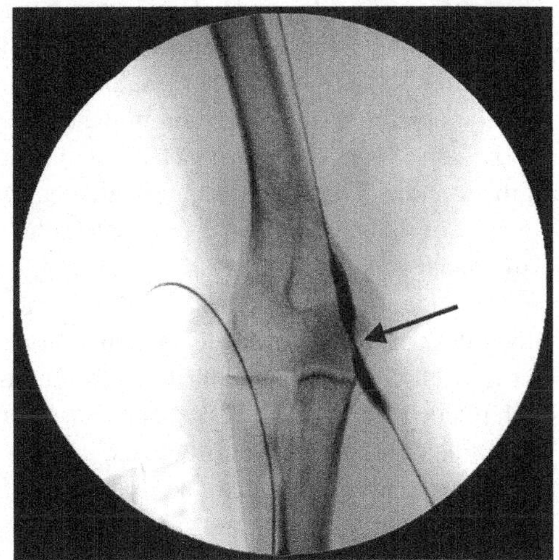

Fig. 48.2 Angioplasty of lesion, *arrow* indicates location of stenosis

with the large collateral vein. It was treated again with angioplasty following which the pain disappeared.

Discussion

This is an unusual case in which the venous stenosis caused neurological symptoms, in this case pain in the distribution of the medial antebrachial cutaneous nerve. The sensory innervation to the forearm is supplied by the medial (anterior and medial aspect), the lateral (lateral aspect), and the posterior antecubital cutaneous nerves. After originating from the medial cord of the brachial plexus, the medial antecubital

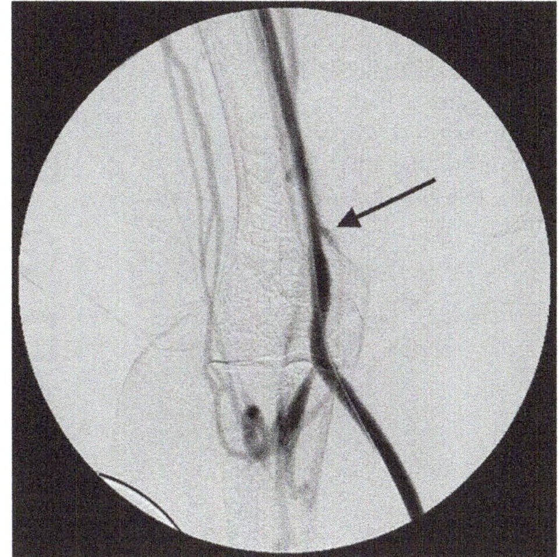

Fig. 48.3 Post-procedure angiogram showing almost complete resolution of collateral vessels, *arrow* indicates location of previous stenosis

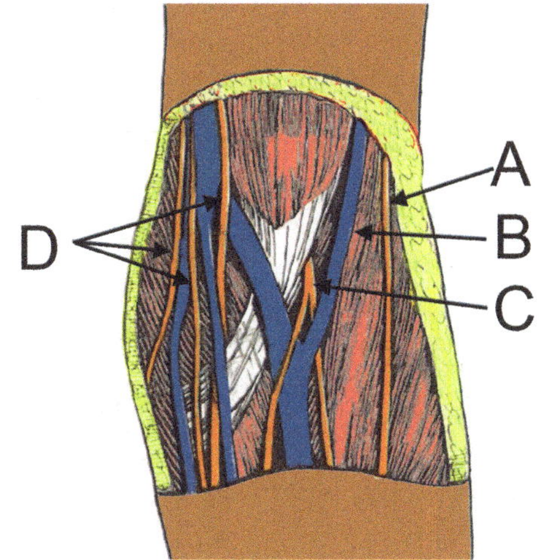

Fig. 48.4 Relationship (right arm) of venous structures and cutaneous nerves to forearm. *A* posterior antebrachial cutaneous nerve, *B* cephalic vein, *C* lateral antebrachial cutaneous nerve, *D* medial antebrachial cutaneous nerve

cutaneous nerve (Figs. 48.3 and 48.4) lies medial to and in close proximity to the basilic vein in the upper arm. In the lower portion of the upper arm, it pierces the brachial fascia and, as it courses downward, lies in close proximity to the median antecubital vein as it enters the forearm. It innervates the skin of the anterior and medial surfaces of the forearm as far down as the wrist (Fig. 48.5).

Pain in the area of an AVG is not an unusual complaint; however, it is generally present only during dialysis or with cannulation. The fact that this patient's pain was continuous and intractable made it different. The response of the pain to the angioplasty suggests that the sensory nerve was being compressed by the engorged collateral veins resulting from the venous stenosis. The recurrence of the symptoms and their subsequent relieve with repeat angioplasty is analogous to meeting Koch's postulates of disease causation.

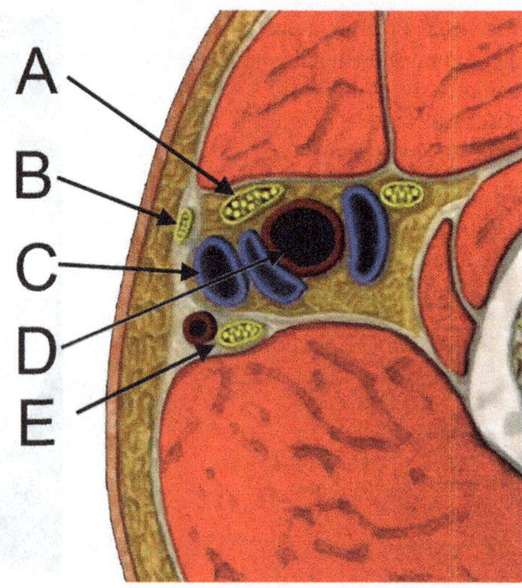

Fig. 48.5 Cross-sectional anatomy at level of the elbow. *A* median nerve, *B* medial antebrachial cutaneous nerve, *C* basilic vein, *D* brachial artery and paired brachial veins, *E* ulnar nerve

The culprit in this case was a collateral vein that originated between the venous end of the graft and the anastomosis-related stenosis. Although it seems as though there should not be a vessel in this area, such a finding is not particularly unusual for patients with an AVG.

The exact incidence of the phenomenon represented in this case is not known; however, there are other situations in which venous congestion secondary to dialysis access-related stenosis has been found to cause serious symptoms not generally thought of as being access related. Seizures have been reported secondary to dialysis access-related central vein stenosis resulting from intracranial venous engorgement [1]. High-pressure glaucoma causing an increasing loss of vision has also been reported related to retinal vein engorgement secondary to dialysis access-related central vein stenosis [2]. Although unusual, these syndromes are important to recognize because of their serious nature.

References

1. Herzig DW, Stemer AB, Bell RS, et al. Neurological sequelae from brachiocephalic vein stenosis. J Neurosurg. 2013;118:1058.
2. Kiernan M, Bhogal M, Wong K, et al. Sight-threatening intraocular pressure due to an upper arm dialysis fistula. Lancet. 2015;386:101.

Part VI
Arterial Interventions

Chapter 49
Directional Atherectomy for Arteriovenous Access Dysfunction

Ali Gardezi and Alexander S. Yevzlin

Case Presentation

A 59-year-old male presented with thrombosis of polytetrafluoroethylene (PTFE) arteriovenous graft (AVG). He had a history of thrombosis of the graft 371 days prior to the current presentation. At that time, a bare-metal stent (BMS) was placed at the stenosed venous anastomosis after thrombectomy. Four months prior to the current presentation, he had a successful angioplasty of a 90% in-stent restenosis.

After the patient was prepped in a usual sterile fashion, a Fogarty balloon and thromboaspiration was used for thrombectomy. After restoration of the flow through the graft, a post-intervention angiogram showed a 90% stenosis of the entire length of the previously placed stent at the venous anastomosis (Fig. 49.1). Angioplasty was performed using a 6 mm × 4 cm balloon inflated to 25 atm for 45 s at the level of the stenosis. Post-interventional angiogram revealed a persistent 80% stenosis. Due to resistant nature, severity, and length of the stenosis, a decision was made to attempt directional atherectomy. The previously placed 6-Fr, 5 cm sheath was replaced by a 7-Fr, 45-cm bendable introducer/sheath. A directional atherectomy (DA) device (Silverhawk, Foxhollow Technologies, Redwood City, CA, USA) was advanced over a 0.014-inch guidewire across the lesion to remove hyperplastic cellular matrix from the lumen of the stented AVG and vein (Fig. 49.2). Atherectomy was performed in four directions by placing the blade of the DA device in dorsal, ventral, lateral, and medial directions turn by turn. The device was then removed. Post-intervention angiogram revealed a 50% residual stenosis. Another balloon angioplasty was done at the level of the stenosis inflating a 6 mm × 4 cm balloon at

A. Gardezi, MBBS
University of Wisconsin-Madison, Madison, WI, USA

A.S. Yevzlin, MD, FASN (✉)
University of Michigan, Ann Arbor, MI, USA
e-mail: asy@medicine.wisc.edu

© Springer International Publishing AG 2017
A.S. Yevzlin et al. (eds.), *Dialysis Access Cases*,
DOI 10.1007/978-3-319-57500-1_49

Fig. 49.1 Post-thrombectomy angiogram revealing a severe, lengthy lesion throughout the previously placed stent at the venous anastomosis

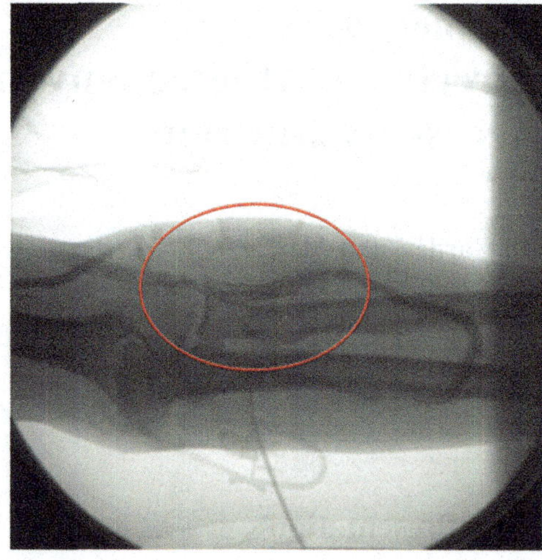

Fig. 49.2 The directional atherectomy device advanced across the lesion to remove hyperplastic cellular matrix

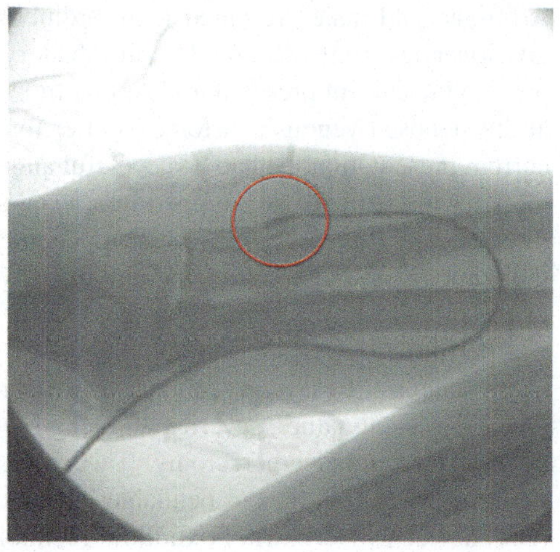

25 atm for 30 s. Post-intervention angiogram revealed the resolution of the in-stent lesion to 15% residual stenosis. Hemostasis was secured using a 2.0 silk suture. There were no complications during the procedure.

Discussion

This case describes the use of DA to treat in-stent restenosis at the venous anastomosis of a thrombosed PTFE AVG. Although it uses a DA device, this procedure is, technically, not an atherectomy. Rather than removing an atheroma, it involves sharp endoluminal dissection to remove the intimal hyperplasia and fibrosis.

Stenosis remains one of the most important complications of dialysis access. In case of AVG, stenosis at venous anastomosis can lead to graft thrombosis [1]. Stent deployment is used in lesions which are either resistant to angioplasty or recur frequently. However, stent placement in veins is not as successful as it is in arteries due to higher rates of in-stent restenosis resulting in lower primary patency [2]. This leads to the dilemma of either placing a new stent within an existing stent or leaving the in-stent restenosis without further intervention [3].

In-stent restenosis has previously been treated with angioplasty alone, angioplasty with another stent placement, or cutting balloon angioplasty. Cutting balloon angioplasty has been used in a variety of vascular lesions [4]. Singer-Jordan et al. described a series of 29 patients undergoing cutting balloon angioplasty on arteriovenous fistulas (AVF). It showed good immediate results in achieving 100% patency, but 6 months primary patency rates were the same as conventional angioplasty [5]. Similarly, a later study by Kariya et al. demonstrated no advantage over conventional angioplasty in treating in-stent restenosis of hemodialysis access [6]. DA is a different approach to the management of restenosis as it removes the fibrosis and neointimal hyperplasia rather than just compressing it against the vessel wall as in conventional or cutting balloon angioplasty.

Several investigators have published their experience of using DA for the treatment of hemodialysis access stenosis. An early case series by Zemel et al. indicated that long-term patency might be superior to percutaneous transluminal angioplasty (PTA) [7]. A subsequent study by Mizumoto et al. showed primary patency rates of 100% at 1 month, 93% at 3 months, 92% at 6 months, and 75% at 12 months [8]. Gray et al. showed that directional atherectomy with or without balloon angioplasty had 83% initial success rates. However, recurrence rates of venous outflow stenosis were similar to balloon angioplasty using the same criterion [9]. DA has not been used for in-stent restenosis before this case report. In fact, the presence of a stent has been described as a relative contraindication for DA due to the risk of perforation and catching the blades of the device on the struts of the stent [7]. Due to these risks, extreme caution is recommended while applying DA in or near a previously deployed stent. Despite these risks, the use of DA device to treat severe and resistant in-stent restenosis is promising, especially so, since it represents an opportunity to study the cells harvested from the lesion.

References

1. Maya ID, Allon M. Outcomes of thrombosed arteriovenous grafts: comparison of stents vs angioplasty. Kidney Int. 2006;69:934–7.
2. Maya ID, Saddekni S, Allon M. Treatment of refractory central vein stenosis in hemodialysis patients with stents. Semin Dial. 2007;20:78–82.
3. Aruny JE, Lewis CA, Cardella JF, Cole PE, Davis A, Drooz AT, Grassi CJ, Gray RJ, Husted JW, Jones MT, TC MC, Meranze SG, Van Moore A, Neithamer CD, Oglevie SB, Omary RA, Patel NH, Rholl KS, Roberts AC, Sacks D, Sanchez O, Silverstein MI, Singh H, Swan TL, Towbin RB. Quality improvement guidelines for percutaneous management of the thrombosed

or dysfunctional dialysis access. Standards of Practice Committee of the Society of Cardiovascular and Interventional Radiology. J Vasc Interv Radiol. 1999;10:491–8.
4. Tsetis D, Morgan R, Belli AM. Cutting balloons for the treatment of vascular stenoses. Eur Radiol. 2006;16:1675–83.
5. Singer-Jordan J, Papura S. Cutting balloon angioplasty for primary treatment of hemodialysis fistula venous stenoses: preliminary results. J Vasc Interv Radiol. 2005;16:25–9.
6. Kariya S, Tanigawa N, Kojima H, Komemushi A, Shomura Y, Shiraishi T, Kawanaka T, Sawada S. Primary patency with cutting balloon angioplasty for different types of hemodialysis access stenosis. Radiology. 2007;243:578–87.
7. Zemel G, Katzen BT, Dake MD, Benenati JF, Lempert TE, Moskowitz L. Directional atherectomy in the treatment of stenotic dialysis access fistulas. J Vasc Interv Radiol. 1990;1:35–8.
8. Mizumoto D, Watanabe Y, Kumon S, Kato H, Haruta Y, Kudo N, Ito K, Ito S, Sakai K, Fukuzawa Y. The treatment of chronic hemodialysis vascular access by directional atherectomy. Nephron. 1996;74:45–52.
9. Gray RJ, Dolmatch BL, Buick MK. Directional atherectomy treatment for hemodialysis access: early results. J Vasc Interv Radiol. 1992;3:497–503.

Chapter 50
Excimer Laser Atherectomy

Alexander S. Yevzlin

Case Presentation

A 55-year-old woman, who was diagnosed with IgG lambda multiple myeloma, was referred to Interventional Nephrology for evaluation of arm swelling on the ipsilateral side of a recently created arteriovenous fistula (AVF). She had been receiving hemodialysis for the past 3 months via right internal jugular vein tunneled catheter. Two prior left upper extremity attempts at AVF placement had failed. A right upper extremity brachial artery to cephalic vein perforator AVF was created several weeks prior to presentation.

After informed consent was obtained, the patient was prepped and draped in the usual sterile fashion overlying the left upper extremity. Two milliliter of subcutaneous lidocaine was administered in the area overlying the AVF. Cannulation was performed with an introducer needle. Guidewire was inserted into the needle. The needle was removed and a 4-French dilator introducer combination was inserted into the lumen of the AVF. The dilator and wire were removed. Introducer was flushed. Contrast was used to define venous anatomy. This revealed a near-total occlusion in the right subclavian vein with no flow of contrast into the SVC around the indwelling tunneled hemodialysis catheter, with the classic "bird's beak" appearance on angiography (Fig. 50.1).

A 0.014 inch guidewire (Ironman, Guidant, Santa Clara, CA) was advanced across the lesion through a 4-Fr catheter. The catheter was attempted to be advanced over the wire unsuccessfully, meeting resistance in the lesion. A 2.5-Fr catheter was advanced to the lesion over the wire but also could not be advanced across successfully due to resistance in the lesion.

A.S. Yevzlin, MD, FASN (✉)
University of Michigan, Ann Arbor, MI, USA
e-mail: asy@medicine.wisc.edu

© Springer International Publishing AG 2017
A.S. Yevzlin et al. (eds.), *Dialysis Access Cases*,
DOI 10.1007/978-3-319-57500-1_50

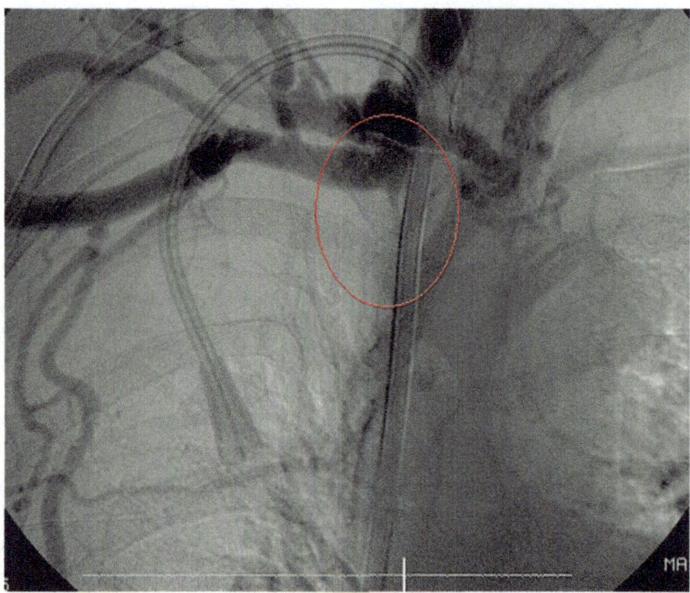

Fig. 50.1 Central venogram revealing near-total occlusion of the right subclavian vein with no flow of contrast into the SVC around the indwelling tunneled hemodialysis catheter and the classic "bird's beak" appearance

Fig. 50.2 Post-laser central venogram showing flow of contrast into the SVC

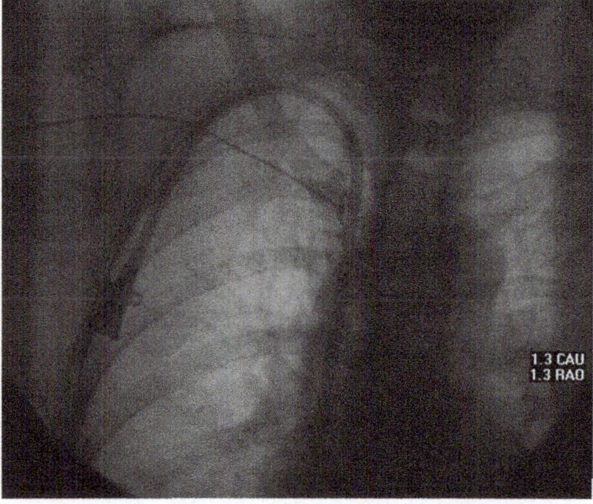

The 4-Fr introducer was exchanged for an 8-Fr. A 2.0 mm laser atherectomy catheter (Turboelite, Spectranetics) was advanced to the level of the lesion over the wire. The laser was engaged at a setting of 40/30 and gradually advanced across the lesion over 15 s duration. Post-laser central venogram revealed flow of contrast into the SVC with 95% residual stenosis (Fig. 50.2). An 8 × 40 mm angioplasty balloon

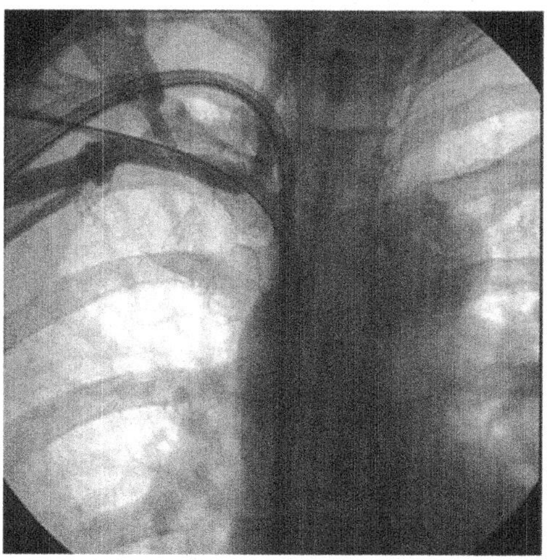

Fig. 50.3 Angiogram of the lesion post-angioplasty

was inserted across the lesion and was inflated to 20 atm for 30 s. Post-angioplasty central venogram revealed good flow of contrast from the subclavian vein into the SVC and right atrium, with a 20% residual stenosis (Fig. 50.3).

Discussion

Excimer laser therapy can be used in the central vasculature of a hemodialysis patient to allow subsequent angioplasty of a 100% stenosis. Excimer (EXCited dIMER) laser technology describes optical amplification that occurs in plasma containing excited dimers with an antibonding electronic ground state [1]. Modern laser technology uses the ultraviolet (UV) spectral region to generate nanosecond pulses of energy. The excimer gain medium is a gas mixture, typically containing a noble gas and a halogen. After stimulated or spontaneous emission, the excimer rapidly dissociates, so that reabsorption of the generated radiation is avoided. This makes it possible to perform photoablation, which is the process by which energy photons cause molecular bond disruption while minimizing thermal damage to the surrounding vascular tissues [2].

We applied the excimer laser technology to the severe neointimal hyperplastic lesion found in the case described above, and for which conventional techniques were not sufficient to obtain a satisfactory result. In occlusive disease where no conventional alternative is available or has been successful, however, laser therapy may be a viable option.

References

1. Ewing JJ. Excimer laser technology development. IEEE J Quantum Electron. 2000;6:1061.
2. Isner JM, Rosenfield K, White CJ, Ramee S, Kearney M, Pieczek A, Langevin RE Jr, Razvi S. In vivo assessment of vascular pathology resulting from laser irradiation. Analysis of 23 patients studied by directional atherectomy immediately after laser angioplasty. Circulation. 1992;85:2185–96.

Chapter 51
Focal Arterial Lesion of the Upper Extremity

Alexander S. Yevzlin

Case Presentation

After informed consent was obtained, the patient was prepped and draped in the usual sterile fashion overlying the left upper extremity. Two millilitre of subcutaneous lidocaine was administered in the area overlying the arteriovenous fistula (AVF). Cannulation was performed with an introducer needle. Guidewire was inserted into the needle. The needle was removed, and a 6 French dilator introducer combination was inserted into the lumen of the AV fistula. The dilator and wire were removed. Introducer was flushed. Contrast was used to define venous anatomy.

A direct arteriogram was then performed which showed normal inflow into the lumen of the AVF from the radial artery, a high bifurcation of radial artery in the upper arm (Fig. 51.1), and a 80% focal lesion at the wrist where a previous access had been surgically created and poor perfusion of the digits.

The catheter was then advanced into the axillary artery, and a direct angiogram was performed, revealing normal flow of the brachial artery but diminutive, diffusely diseased ulnar and radial vessels with no flow into the hand (Fig. 51.2).

Two thousand units of heparin were administered i.v. A 180 cm .014 inch guidewire crossing the radial arterial lesion. Angioplasty balloon 3 × 30 mm was then advanced to the distal lesion where it was inflated for 5 s at 8 atmospheres (Fig. 51.3). Post-intervention direct arteriogram revealed a 0% residual stenosis and much better palmar arch (Fig. 51.4).

A.S. Yevzlin, MD, FASN (✉)
University of Michigan, Ann Arbor, MI, USA
e-mail: asy@medicine.wisc.edu

© Springer International Publishing AG 2017
A.S. Yevzlin et al. (eds.), *Dialysis Access Cases*,
DOI 10.1007/978-3-319-57500-1_51

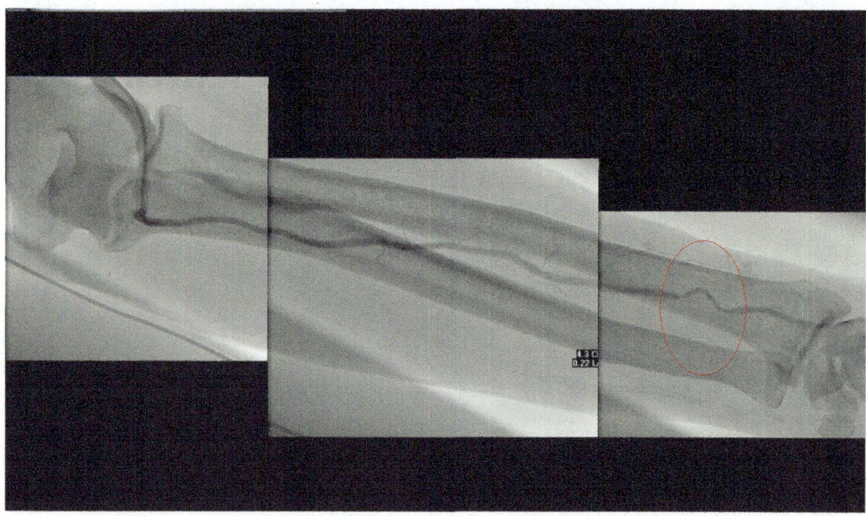

Fig. 51.1 Direct arteriogram showing normal inflow into the lumen of the arteriovenous fistula from the radial artery, a high bifurcation of radial artery in the upper arm, and an 80% focal lesion at the wrist where a previous access had been surgically created and poor perfusion of the digits

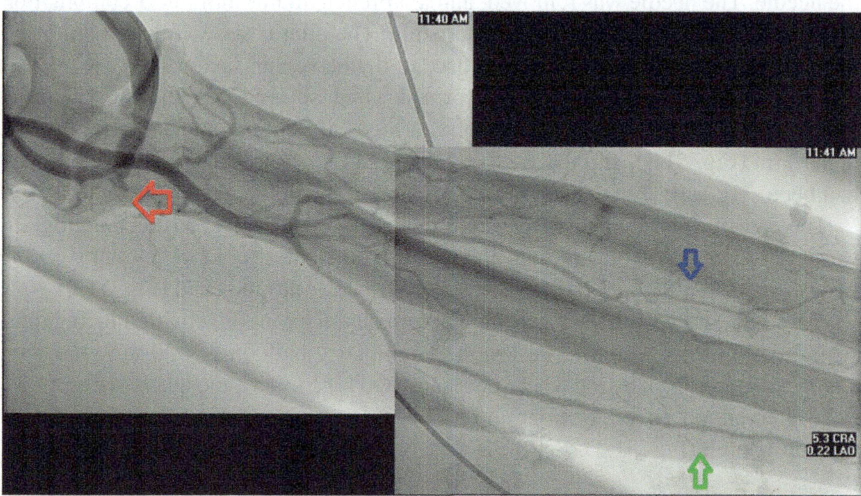

Fig. 51.2 Direct angiogram of the axillary artery, revealing normal flow of the brachial artery but diminutive, diffusely diseased ulnar (*green arrow*) and radial (*blue arrow*) vessels with no flow into the hand. *Red arrow* shows arteriovenous fistula anastomosis with high-bifurcation radial artery

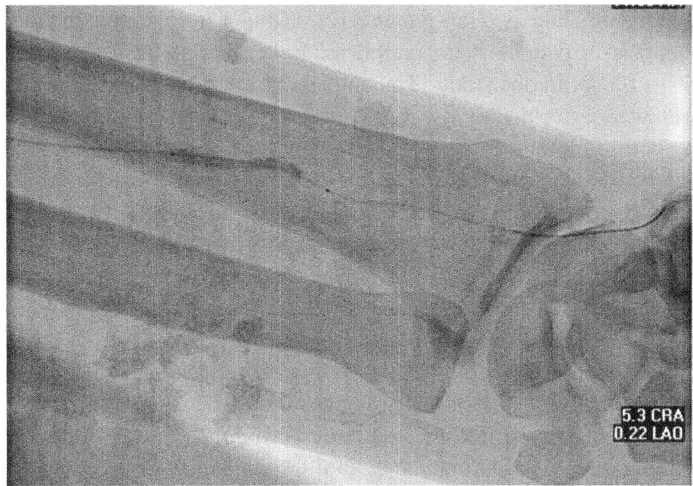

Fig. 51.3 A 3 × 30 mm angioplasty balloon intervention over 0.14 inch wire

Fig. 51.4 Post-intervention direct arteriogram revealed a 0% residual stenosis

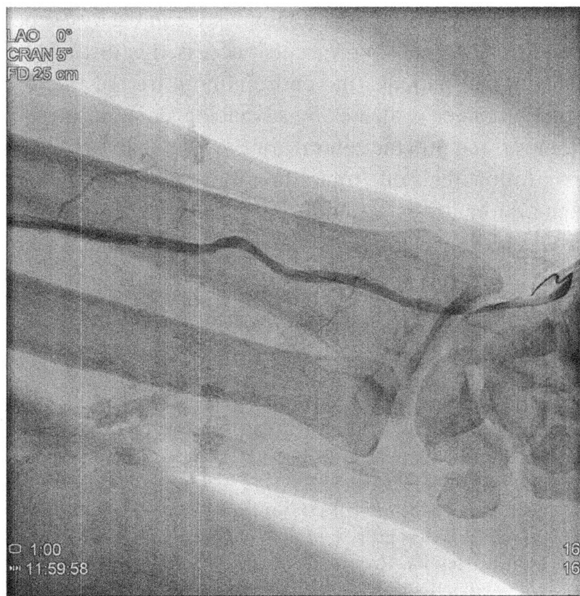

Discussion

Recent data have emphasized that significant (≥50%) arterial stenoses are commonly seen in dialysis patients presenting with symptoms of hand ischemia [1]. These lesions can occur anywhere within the arteries of the upper extremities including the proximal arteries and have been demonstrated to cause peripheral ischemia in haemodialysis (HD) patients [2]. Using arteriography, the incidence of arterial stenosis in patients with peripheral ischemia has been reported to range from 62% to 100%. In one study [3], complete arteriography from the aortic to the palmar arch was performed to assess the presence of arterial stenosis in HD patients presenting with symptoms of peripheral ischemia ($n = 13$). It was found that 62% of the 13 patients referred for the evaluation of symptoms of steal syndrome demonstrated a significant ($\geq 50\%$) arterial stenosis. In another report [4], stenosis in the inflow circulation was found in 100% of the patients who underwent complete arteriography ($n = 5$).

Arteriography is the most important tool in making the diagnosis as well as developing a treatment strategy for hand ischemia [2]. As steal from the distal vessels can be expected, arteriography should be performed with and without occlusion of the arteriovenous access. Optimally, images should be recorded from the aortic to the palmar arch so that the lesions in the more proximal arteries will not be missed [5]. Recorded images should then be carefully evaluated for the presence of occlusive arterial disease. Both the femoral artery and cannulating the access in a retrograde direction and advancing a diagnostic catheter into the aortic arch area can be used to perform angiography and angioplasty.

Percutaneous balloon angioplasty can easily be performed to treat arterial stenotic lesions causing vascular access dysfunction or hand ischemia. In a great majority of patients, the entire arterial tree of the extremity including the central arteries can be evaluated by advancing a diagnostic catheter across the arterial anastomosis and into the central arteries. Navigation of the balloon to the site of stenosis can follow the same route. In a minority of patients, direct arterial puncture of the femoral or upper extremity arteries might be needed to evaluate and manage the arterial lesions [1].

References

1. Yevzlin AS, Schoenkerman AB, Gimelli G, Asif A. Arterial interventions in arteriovenous access and chronic kidney disease: a role for interventional nephrologists. Semin Dial. 2009;22(5):545–56.
2. Asif A, Leon C, Merrill D, Bhimani B, Ellis R, Ladino M, Gadalean FN. Arterial steal syndrome: a modest proposal for an old paradigm. Am J Kidney Dis. 2006;48:88–97.
3. Valji K, Hye RJ, Roberts AC, Oglevie SB, Ziegler T, Bookstein JJ. Hand ischemia in patients with hemodialysis access grafts: angiographic diagnosis and treatment. Radiology. 1995;196:697–701.
4. Tordoir JHM, Dammers R, van der Sande FM. Upper extremity ischemia and hemodialysis vascular access. Eur J Vasc Endovasc Surg. 2004;27:1–5.
5. Guerra A, Raynaud A, Beyssen B, Pagny JY, Sapoval M, Angel C. Arterial percutaneous angioplasty in upper limbs with vascular access devices for haemodialysis. Nephrol Dial Transplant. 2002;17:843–51.

Chapter 52
Brachial Artery Stenosis

Gerald A. Beathard

Case Presentation

The patient was a 63-year-old diabetic male who was dialyzed for 3 years and had a brachial-basilic arteriovenous fistula (AVF) in his left arm. He was referred to the dialysis access clinic with stage III dialysis access steal syndrome (DASS). The patient had a history of left hand problems that started approximately 3 months earlier with just a cold hand, progressing to pain during dialysis and eventually pain when not on dialysis.

Physical examination revealed that the patient had a large brachial-cephalic AVF, which collapsed with arm elevation. The right hand was cold in comparison to the left. The radial pulse was not palpable and did not change with manual occlusion of the AVF. A bruit was detectable with Doppler and improved only slightly with AVF occlusion. Ultrasound evaluation revealed that flow in the radial artery was retrograde during diastole. Blood flow in the brachial artery measured above the anastomosis was 670 mL/min.

The patient was prepped and draped. The AVF was cannulated in a retrograde direction, and an angiogram was performed that showed no venous abnormality. There was no flow of radiocontrast into the radial artery with the AVF open; the flow was minimal with the occlusion. An arteriogram was performed that showed two areas of stenosis in the brachial artery (Fig. 52.1). The radial artery was not stenotic but did show irregularity in the lumen.

The brachial artery lesions were dilated with a 6 × 4 balloon with good result (Fig. 52.2). Following the treatment, there was minimal flow in the radial artery with the AVF open and good flow with the occlusion. The patient's hand was warm

G.A. Beathard, MD, PhD (✉)
Lifetime Vascular Access, University of Texas Medical Branch,
5135 Holly Terrace, Houston, TX 77056, USA
e-mail: gbeathard@msn.com

© Springer International Publishing AG 2017
A.S. Yevzlin et al. (eds.), *Dialysis Access Cases*,
DOI 10.1007/978-3-319-57500-1_52

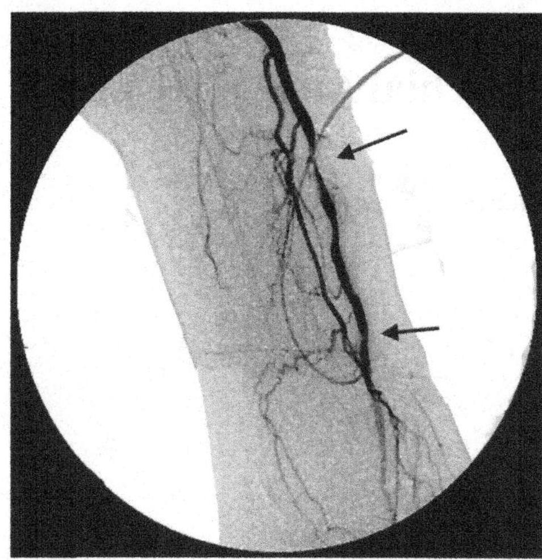

Fig. 52.1 Angiogram of brachial artery. *Arrows* indicate stenotic lesions

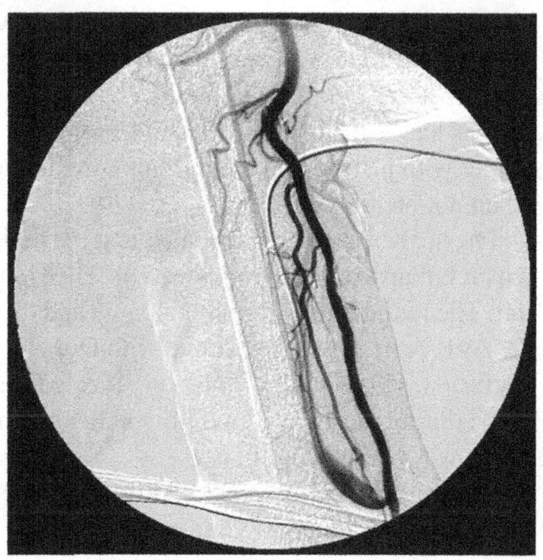

Fig. 52.2 Postangioplasty angiogram showing resolution of lesions

following the procedure. Flow in the brachial artery measured above the anastomosis was 1080 mL/min. The following day, dialysis was accomplished without any hand pain.

Discussion

This dialysis patient with hand ischemia illustrates the importance of evaluating the arterial blood flow to the AVF. Dialysis access steal syndrome (DASS) is seen primarily in AVFs, and the incidence increases with time. This is because of two issues. Access blood flow increases with time and problems such as arteriosclerosis that can lead to hand ischemia in the patient with an access advance with time.

In this case, retrograde flow was present as it is in most patients with an AVF; however, this is a physiologic phenomenon (physiologic steal) and as such was not involved in the pathogenesis of hand ischemia in this case [1]. It should be noted that unlike the retrograde flow that is seen with a radial artery based AVF, the retrograde flow with the brachial artery AVF is actually to-and-fro; it is antegrade during systole and retrograde in diastole.

When evidence for the existence of DASS is detected or even suspected, the patient should be immediately subjected to a complete evaluation. This should include a measurement of the access blood flow and a complete angiogram of the access and its feeding artery, including distal runoff. Proximal arterial stenosis affecting the feeding artery is the cause of access dysfunction in a significant number of patients with access-related ischemia [2, 3] as was the case here.

References

1. Beathard GA, Spergel LM. Hand ischemia associated with dialysis vascular access: an individualized access flow-based approach to therapy. Semin Dial. 2013;26:287.
2. DeCaprio JD, Valentine RJ, Kakish HB, et al. Steal syndrome complicating hemodialysis access. Cardiovasc Surg. 1997;5:648.
3. Asif A, Leon C, Merrill D, et al. Arterial steal syndrome: a modest proposal for an old paradigm. Am J Kidney Dis. 2006;48:88.

Part VII
Hand Ischemia

Chapter 53
True Steal Syndrome

Eduardo Rodriguez and Karl A. Illig

Case Presentation

A 41-year-old man with a left upper-arm brachiocephalic arteriovenous fistula (AVF) presents to the emergency room 6 months after dialysis access creation with complaints of worsening shortness of breath and a nonhealing wound on the left index finger. Physical examination revealed a brisk thrill over the AVF, nonpalpable distal pulses in the left hand, and evidence of dry gangrene at the distal left index finger. A plain X-ray of the hand was consistent with osteomyelitis involving the base of the distal phalanx of the second digit and head of the middle phalanx of the second digit. Noninvasive studies on the left upper extremity revealed a left finger pressure of 38 mmHg that improved to 110–120 mmHg, an absence of evidence of inflow disease, a fistula volume flow of 1200 mL/min, a biphasic signal in the radial artery, and an improvement of flow and signal with AVF compression. These findings were suggestive of ischemic tissue loss secondary to "true" steal syndrome.

The patient was taken to the operating room for distal revascularization and interval ligation (DRIL). Exposure was obtained in the axillary artery as high as possible in the axilla and the brachial artery just distal to the antecubital fossa at its bifurcation. The greater saphenous vein was harvested from the left thigh. Proximal anastomosis was constructed, and the vein was then tunneled in a reversed fashion. Distal anastomosis was constructed to the brachial bifurcation with ligation performed between the distal anastomosis and the fistula itself completing the procedure. Handheld Doppler revealed excellent flow in the DRIL, which was not

E. Rodriguez, MD (✉)
Memorial Regional Hospital, Department of Vascular Surgery, 1150 N. 35th Ave,
Ste 460, Hollywood, FL 33160, USA
e-mail: erodriguezzoppi@mhs.net

K.A. Illig, MD
Division of Vascular Surgery, Department of Surgery, University of South
Florida Morsani College of Medicine, Tampa, FL, USA

© Springer International Publishing AG 2017
A.S. Yevzlin et al. (eds.), *Dialysis Access Cases*,
DOI 10.1007/978-3-319-57500-1_53

appreciably affected by occlusion of the patent fistula. A left second finger amputation was performed at the same time.

Discussion

Access-related hand ischemia or steal syndrome occurs secondary to low access tract resistance, causing reversal of blood flow in the arterial outflow tract toward the access and away from the hand. This reversal of flow following arteriovenous (AV) access creation is a very common hemodynamic finding. Studies have shown reversal of flow in 73% of autogenous access (AA) and in 91% of prosthetic AV accesses [1]. Nonetheless, a symptomatic steal syndrome is present in less than 10%. Clinical symptoms vary and range from mild ischemia to tissue loss. Steal can be graded based on presentation [2]:

Grade 0: No steal
Grade 1: Mild—cool extremity, few symptoms, flow augmentation with access
     occlusion
Grade 2: Moderate—intermittent ischemia only during dialysis, claudication
Grade 3: Severe—ischemic pain at rest, tissue loss

Patients with clinically symptomatic arterial steal should be evaluated with color duplex ultrasound and digital pneumatic plethysmography. The duplex will conform retrograde flow in the outflow artery distal to the access and the volume flow in the access. The digital pneumatic plethysmography suggests true steal when there is evidence of flat digital waveforms in the affected extremity, with a return of pulsatile waveforms with access compression [3]. In the presence of proximal arterial stenosis (inflow disease), waveforms would remain flat despite access compression. The next step in diagnosis should include an arteriogram to identify any inflow stenosis that could be treated with endovascular interventions, including angioplasty and stenting. The arteriogram would also demonstrate which outflow vessel appears to be the dominant one supplying the forearm and hand.

If digital pressures do not normalize with compression of the access, hand ischemia is most likely due to outflow arterial disease in the arterial tree distal to the access anastomosis. Ligation of the access will instantly resolve the symptoms, but it leaves the patient without a dialysis access. However, if distal arterial circulation cannot be improved, ligation of the fistula may be the only treatment option. In patients with a forearm access and a patent palmer arch, the distal outflow artery may be ligated, but this approach does not restore possible required inflow to the hand.

The volume flow, high or low, through the access can be used to suggest therapeutic options to address steal in the setting of distal arterial disease. Figure 53.1 summarizes the diagnosis and management of true steal syndrome.

In the setting of tissue loss or normal flow access, the best option is DRIL [4]. This procedure consists of ligation of the arterial outflow tract just distal to the

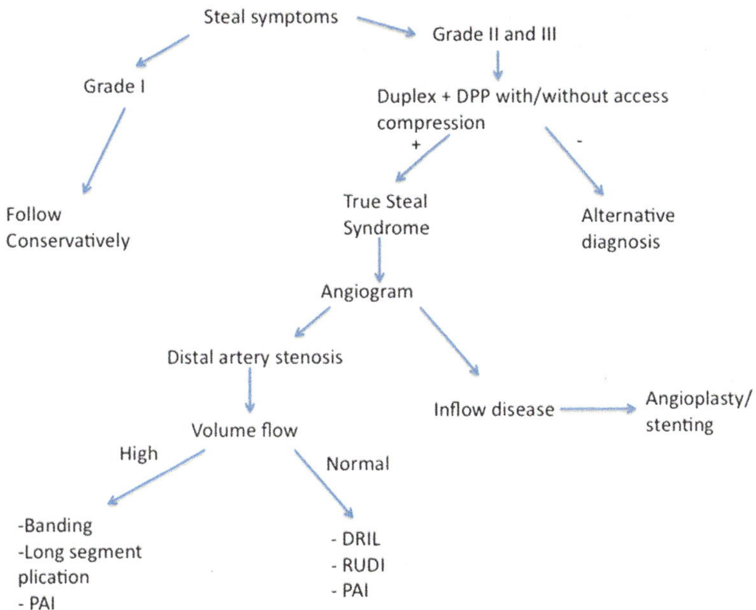

Fig. 53.1 Algorithm for management of true steal syndrome. *DPP* digital pneumatic plethysmography, *DRIL* distal revascularization and interval ligation, *RUDI* revision using distal inflow, *PAI* proximalization of the arterial inflow

takeoff of the access, followed by placement of a bypass originating from the inflow vessel in a location as proximal as possible to the arterial anastomosis. The target vessel of the bypass should be the dominant artery distal to the area of ligation. It is important to locate the origin of the bypass graft high and away from the inflow anastomosis of the access to avoid the pressure sink in the inflow vessel, obviating the need to approach the artery through a reoperative field and adding resistance to the fistula with the longer length [5, 6]. This procedure eliminates reversal of flow in the distal outflow artery while keeping the access patent and perfusion pressures to the hand constant [7]. The main disadvantage of DRIL is dependence of the hand on the bypass graft; however, most studies have shown access patency above 80% at 1 year, with 80–90% bypass graft patency at 4 years [4, 7, 8].

An alternative procedure to treat steal syndrome with normal access flow is revision using distal inflow (RUDI), which relocates the access anastomosis to a smaller, more distal, artery [9]. This results in antegrade flow in the distal artery and decreased size of the AV anastomosis. The main advantage of RUDI is that distal perfusion is maintained through the native arterial system; however, its disadvantage is a higher risk of access thrombosis.

Proximalization of the arterial inflow (PAI) is a strategy that can be applied for both low-flow and high-flow steal. This technique consists of relocation of the AV anastomosis more proximally on the inflow artery [10]. The hemodynamic effect is enhanced distal perfusion due to a smaller pressure drop from the AV access.

In cases of high-flow steal, options include flow restriction procedures like banding or long-segment plication. Banding of the access tract creates a stenosis that increases the resistance to the blood flow in the access and reverses arterial flow in the distal artery [11]. However, it is difficult to judge the degree of stenosis required to alleviate the steal without causing thrombosis of the access. These procedures are discussed in other chapters.

References

1. Kwun KB. Hemodynamic evaluation of angioaccess procedures for hemodialysis. Vasc Surg. 1979;13:170–7.
2. Sidawy AN, et al. Recommended standards for reports dealing with arteriovenous hemodialysis accesses. J Vasc Surg. 2002;35:603–10.
3. Mattson WJ. Recognition and treatment of vascular steal secondary to hemodialysis prostheses. Am J Surg. 1987;154:198–201.
4. Schanzer H, et al. Treatment of angioaccess-induced ischemia by revascularization. J Vasc Surg. 1992;16:861–6.
5. Illig KA, et al. Hemodynamics of distal revascularization–interval ligation. Ann Vasc Surg. 2005;19:199–207.
6. Wixon CL, et al. Understanding strategies for the treatment of ischemic steal syndrome after hemodialysis access. J Am Coll Surg. 2000;191:301–10.
7. Berman SS, et al. Distal revascularization–interval ligation for limb salvage and maintenance of dialysis access in ischemic steal syndrome. J Vasc Surg. 1997;26:393–404.
8. Knox RC, et al. Distal revascularization–interval ligation: a durable and effective treatment for ischemic steal syndrome after hemodialysis. J Vasc Surg. 2002;36:250–6.
9. Minion DJ, et al. Revision using distal inflow: a novel approach to dialysis-associated steal syndrome. Ann Vasc Surg. 2005;19:625–8.
10. Zanow J, et al. Proximalization of the arterial inflow: a new technique to treat access-related ischemia. J Vasc Surg. 2006;43:1216–21.
11. Goel N, et al. Minimally invasive limited ligation endoluminal-assisted revision (MILLER) for treatment of dialysis access-associated steal syndrome. Kidney Int. 2006;70:765–70.

Chapter 54
Dialysis Access Steal Syndrome (DASS) in a Patient with Distal Revascularization and Interval Ligation (DRIL)

Gerald A. Beathard

Case Presentation

The patient was a 35-year-old type 1 diabetic who had been on dialysis for 6 years using a brachial-basilic fistula. Four years earlier, the patient had developed dialysis access steal syndrome (DASS) and had a distal revascularization and interval ligation (DRIL) procedure performed at that time. Over the past few months, the patient has had ipsilateral hand pain on dialysis and recently has started having hand pain and breast. The patient was referred to the dialysis access center for evaluation of this hand pain.

Examination the patient revealed that he had a brachial-basilic fistula in his left upper arm. The fistula was functioning well, pulse augmentation was good, and it was not hyper-pulsatile. The left hand was cold in comparison with the right hand. Neither a radial nor ulnar pulse could be palpated.

After the patient was pulsed and raped prepped and draped, an angiogram was performed. This showed the anatomy of the DRIL procedure that had been previously done. An area of marked stenosis was present at the arterial anastomosis of the graft used for the distal revascularization (Fig. 54.1a, b). No radiocontrast flow to the forearm was noted (Fig. 54.2). After a guidewire was advanced through the interval ligation, the stenotic site was dilated with a 6 × 4 angioplasty balloon (Fig. 54.3). Afterward, blood flow to the forearm was restored although diffuse arteriopathy was obvious (Figs. 54.4 and 54.5).

G.A. Beathard, MD, PhD (✉)
Lifetime Vascular Access, University of Texas Medical Branch,
5135 Holly Terrace, Houston, TX 77056, USA
e-mail: gbeathard@msn.com

© Springer International Publishing AG 2017
A.S. Yevzlin et al. (eds.), *Dialysis Access Cases*,
DOI 10.1007/978-3-319-57500-1_54

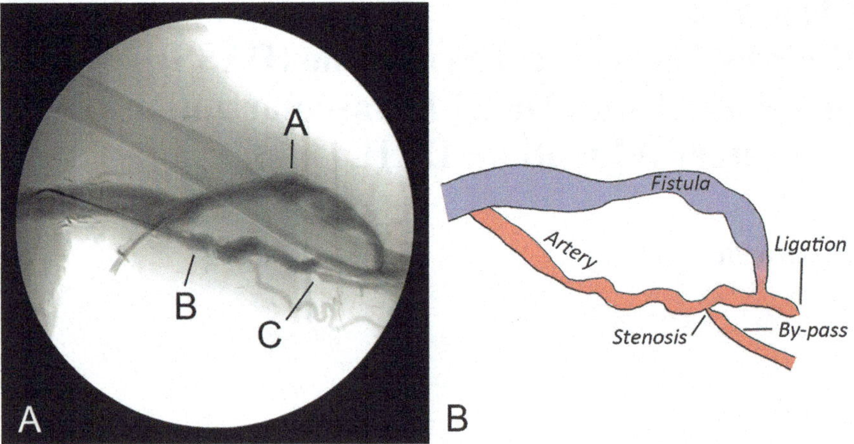

Fig. 54.1 (**a**) Initial angiogram of fistula. *A* fistula, *B* brachial artery, *C* bypass for distal revascularization, note stenosis. (**b**) Drawing of angiogram shown in Fig. 54.1A

Fig. 54.2 Angiogram showing lower portion of distal revascularization and interval ligation (DRIL) anatomy and no radiocontrast passing to the forearm. *Arrow* site of stenosis, *A* site of interval ligation

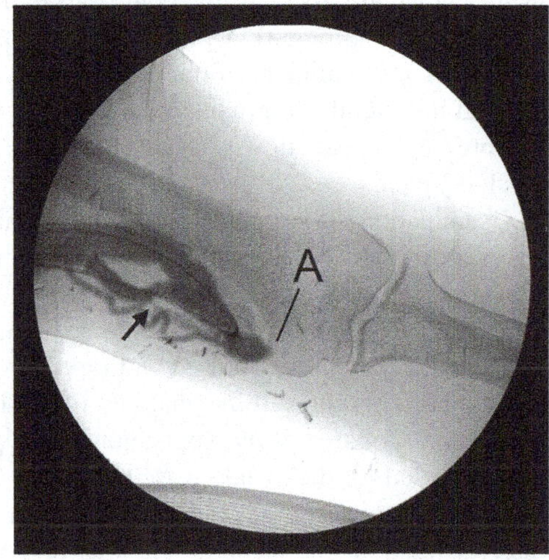

Discussion

Dialysis access steal syndrome resulting in hand ischemia is a complication that can occur with an arteriovenous access. It is more common with an arteriovenous fistula (AVF) than with an arteriovenous graft (AVG) and more common with an AVF associated with the brachial artery. DRIL is one of the procedures that has been used to treat this problem. There are two steps to this procedure [1]. First, the artery distal to the arterial anastomosis is ligated. This interrupts retrograde diastolic inflow into the access, the pathophysiologic principle of steal. Second, a bypass is created connecting the brachial artery above the anastomosis with one of the forearm arteries

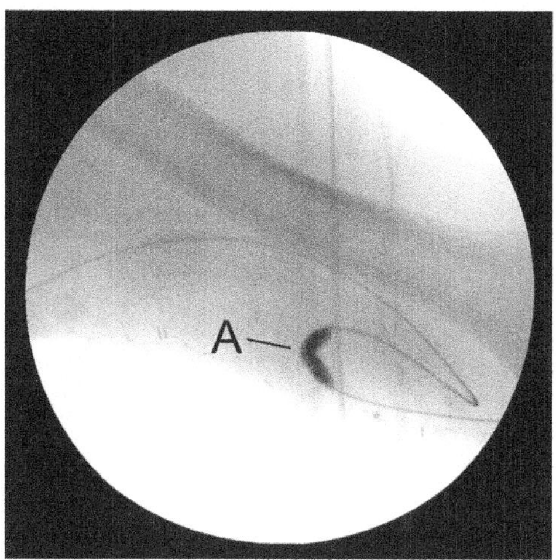

Fig. 54.3 Angioplasty of stenotic lesion. *A* angioplasty balloon

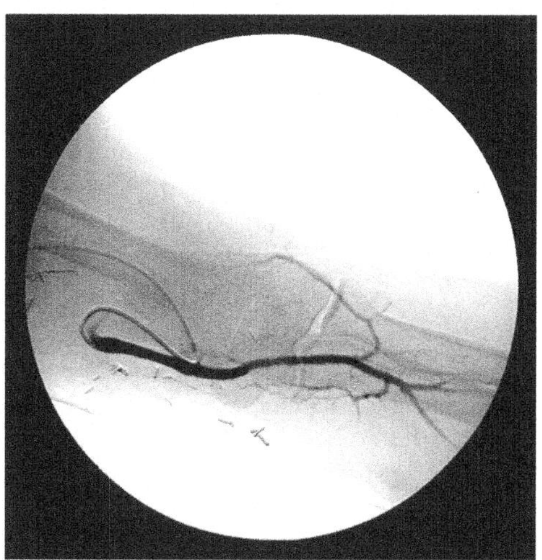

Fig. 54.4 Angiogram of upper forearm

distal to the ligation. Conceptually, DRIL provides an added low-resistance collateral artery, which reduces the total peripheral resistance in the distal extremity. This serves to increase the total peripheral perfusion to the limb. Additionally, it does this while simultaneously blocking retrograde flow to the access, which is a major contributor to the pathophysiology resulting in hand ischemia.

A major concern with the DRIL procedure is related to the bypass upon which the distal circulation is dependent. This can develop problems as was noted in this case resulting in a recurrence of hand ischemia. Because of concerns related to this issue, the DRIL procedure has essentially been abandoned in favor of a procedure referred to as proximalization of the arterial inflow (PAI) in many surgical practices. In this procedure, a small-caliber (4 or 5 mm) expanded polytetrafluoroethylene (ePTFE) graft is connected to the arterial end of the fistula and extended up to a new

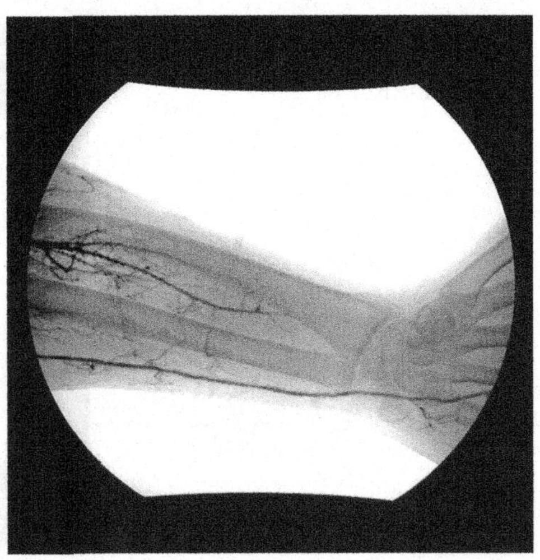

Fig. 54.5 Angiogram of lower forearm, note diffuse arteriopathy

arterial anastomosis with the axillary artery. The original arterial anastomosis is closed, and there is no ligation of blood flow to the lower arm [2, 3]. Another approach has been to perform only the distal revascularization (DL) portion of the DRIL and omit the interval ligation (IL). These two procedures are very similar, differing only in the distance proximal to the original arterial anastomosis that is used for the new inflow to the fistula.

An additionall point that should be made concerning this case relates to the distance between the AVF anastomosis and the arterial anastomosis of the revascularization by=pass. It is recommended that this be at least 10 cm; however as can be seen in Fig. 54.1A above, the distance is much less than this. This problem would make the patient more suseptible to a recurrence of hand ischemia.

References

1. Schanzer H, Schwartz M, Harrington E, Haimov M. Treatment of ischemia due to "steal" by arteriovenous fistula with distal artery ligation and revascularization. J Vasc Surg. 1988;7(6):770–3.
2. Zanow J, Petzold K, Petzold M, Krueger U, Scholz H. Flow reduction in high-flow arteriovenous access using intraoperative flow monitoring. J Vasc Surg. 2006;44(6):1273–8.
3. Thermann F, Wollert U. Proximalization of the arterial inflow: new treatment of choice in patients with advanced dialysis shunt-associated steal syndrome? Ann Vasc Surg. 2009;23(4):485–90.

Chapter 55
Dialysis Access Steal Syndrome Following Percutaneous Transluminal Angioplasty in a Radial-Cephalic Fistula

Gerald A. Beathard

Case Presentation

The patient was a 64-year-old male diabetic who had been receiving dialysis treatments for 2 years. He was referred to the vascular access center because of poor flow in his left radial-cephalic fistula. At that time the patient did admit to have mild pain in his ipsilateral hand during dialysis. After he was prepped and draped, and angiogram was performed which showed the presence of juxta-anastomotic stenosis. This was successfully treated with angioplasty, and the patient was referred back returned to the dialysis clinic where his fistula blood flow was found to be much improved and his fistula was easier to cannulate.

One month later, the patient was referred sent back to the vascular access center with a complaint of worsening hand pain both at dialysis and at rest. Examination revealed that his ipsilateral hand was cold in comparison to the opposite hand, there was dry gangrene of the tip of the ring finger, and the radial pulse was not detectable.

After the patient was prepped and draped, the arterial circulation associated with the radial-cephalic fistula was evaluated from the aortic arch down through the palmar arch and was found to be relatively normal. Radiocontrast forcibly injected into the distal fistula filled the distal radial artery, but immediately flowed retrograde back into the fistula. Radiocontrast injected with the fistula manually occluded flowed into the palmar arch.

A decision was made to occlude the distal radial artery in order to prevent retrograde flow from the palmar arch. This was accomplished using embolization coils. Following this procedure, the patient's hand was appreciably warmer. The patient

G.A. Beathard, MD, PhD (✉)
Lifetime Vascular Access, University of Texas Medical Branch,
5135 Holly Terrace, Houston, TX 77056, USA
e-mail: gbeathard@msn.com

© Springer International Publishing AG 2017
A.S. Yevzlin et al. (eds.), *Dialysis Access Cases*,
DOI 10.1007/978-3-319-57500-1_55

was referred back to the dialysis clinic where his dialysis continued in an uncomplicated fashion. Follow-up at 4 weeks revealed no further hand pain and evidence of healing of the ring finger.

Discussion

It is important to understand the pathophysiology associated with dialysis access steal syndrome (DASS). When an arteriovenous access is placed in an extremity, a complex and unique physiological state is established due to the fact that blood flow to the extremity must now supply the access as well as the hand. This "hand/vascular access complex" consists of a proximal artery feeding into two competing circuits connected in series – the arteriovenous access and the peripheral vascular bed upon which perfusion of the hand is dependent. The former is a low-resistance pathway and is located proximally; the latter is downstream and is characterized by high resistance. Also contributing to this complex are collateral arteries that bypass the access circuit to feed the periphery directly. Changes occur in each of these components during normal fistula development, and each plays a role in the pathophysiology of DASS [1].

In most instances, the effects of the steal phenomenon are offset by other components of the hand/access complex. It is only when these compensatory mechanisms are inadequate that ischemia occurs. The location of the arterial inflow for the fistula is an important determining factor for the development of DASS. A brachial artery asscoiated AVF is more likely to cause problems than radial based one [2, 3].

Changes within the fistula and its draining veins can also mitigate the development of DASS. With the development of a venous stenosis within the fistula or its drainage, resistance in the access circuit increases resulting in a decrease in the steal phenomenon. This can ameliorate or prevent the development of ischemic symptoms. This situation is generally appreciated only when DASS develops or worsens immediately after a successful angioplasty relieves the stenosis as was the case with this patient.

This case represents a situation in which the patient was probably already having mild symptoms of DASS prior to the angioplasty performed for the poor access blood flow since it was noted at that time that he was having mild pain in his hand during dialysis. With treatment of the stenotic lesion, the resistance to blood flow in the fistula was decreased even further resulting in a worsening of the steal. This was severe enough to result in serious hand ischemia and digital tissue loss (Figs. 55.1, 55.2, 55.3, 55.4, 55.5, and 55.6).

The basic problem in DASS associated with a wrist (radial-cephalic) AVF is retrograde flow in the distal radial artery. The problem can be successfully treated using an embolization coil as was done in this case or by surgical ligation to interrupt the retrograde flow. Prior to perfoming this procedure, it is important that the integrity of the ulnar artery be confomirmed.

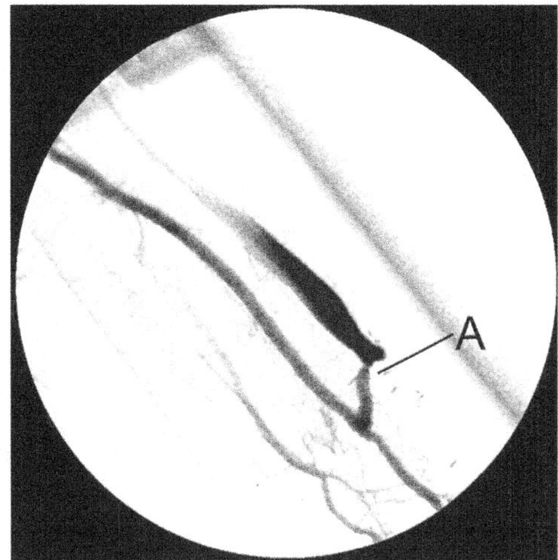

Fig. 55.1 Initial angiogram showing juxta-anastomotic stenosis. *A* site of stenosis

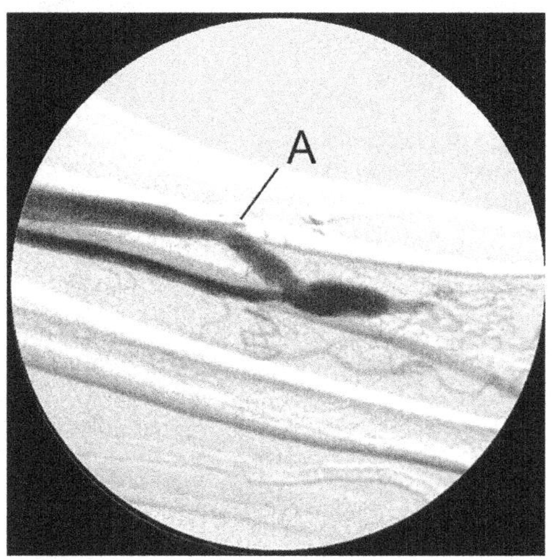

Fig. 55.2 Post-procedure angiogram showing results of angioplasty. *A* site of previous stenosis

While not a common complication, any instance in which a patient begins to complain of ipsilateral hand pain following an angioplasty should raise the question of the emergence of DASS. An additional point illustrated by this case is the fact that one should be very judicious in performing an angioplasty that will increase fistula blood flow in any patient already suspected of having any degree of compromise to their hand circulation. In these cases, if the fistula blood flow is not sufficient to permit effective dialysis, surgical consultation should be sought to devise a plan to accomplish the dual goals of an access capable of permitting adequate dialysis and preserving the function of the hand.

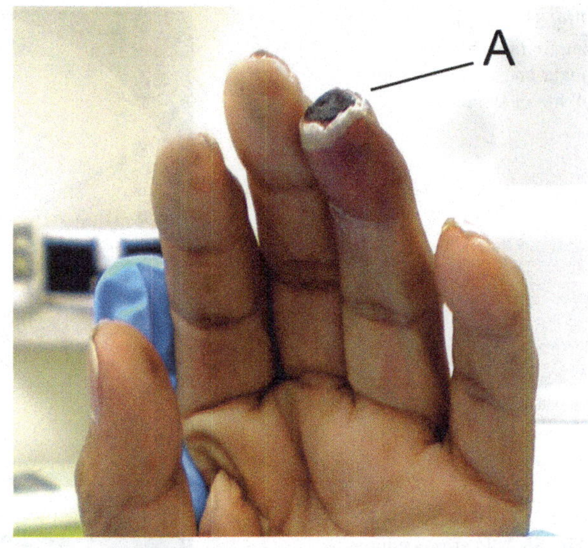

Fig. 55.3 Appearance of patient's ischemic hand showing drag and green tissue loss at the tip of the ring finger

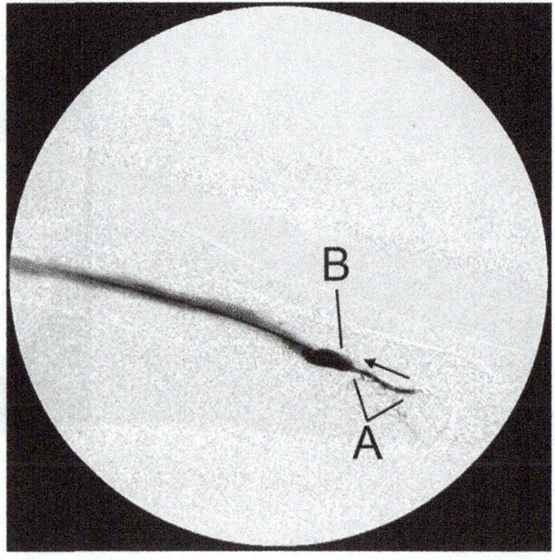

Fig. 55.4 Injection of radiocontrast into distal fistula; *arrowed* indicates direction of flow in radial artery. *A* radial artery, *B* arterial anastomosis of fistula

Fig. 55.5 Injection of radiocontrast into distal fistula with fistula occluded proximally; *arrowed* indicates direction of flow in radial artery. *A* radial artery, *B* arterial anastomosis of fistula

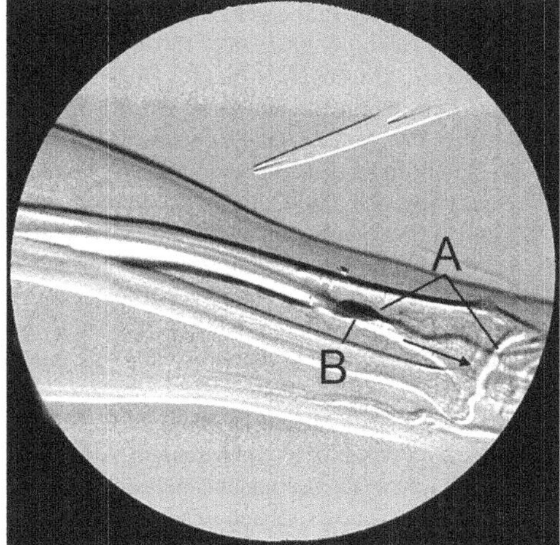

Fig. 55.6 Injection of radiocontrast following placement of embolization coils in radial artery below fistula. *A* radial artery, *B* arterial anastomosis of fistula, *C* embolization coils in radial artery

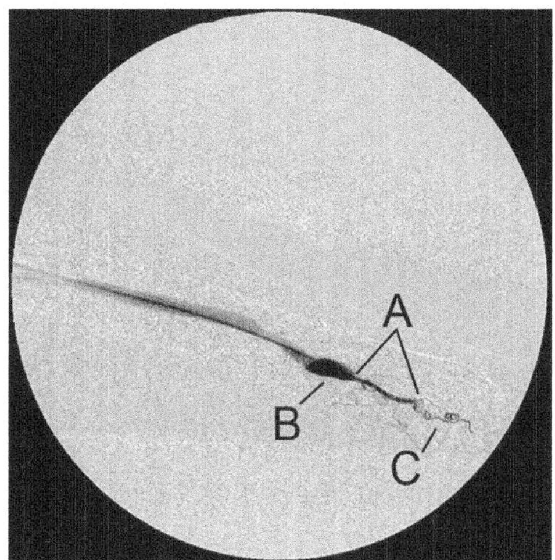

References

1. Beathard GA, Spergel LM. Hand ischemia associated with dialysis vascular access: an individualized access flow-based approach to therapy. Semin Dial. 2013;26(3):287–314.
2. Spergel LM, Ravani P, Roy-Chaudhury P, Asif A, Besarab A. Surgical salvage of the autogenous arteriovenous fistula (AVF). J Nephrol. 2007;20(4):388–98.
3. Scheltinga MR, van Hoek F, Bruijninckx CM. Time of onset in haemodialysis access-induced distal ischaemia (HAIDI) is related to the access type. Nephrol Dial Transplant. 2009;24(10):3198–204.
4. Mitchell GF, Parise H, Vita JA, Larson MG, Warner E, Keaney JF Jr, et al. Local shear stress and brachial artery flow-mediated dilation: the Framingham Heart Study. Hypertension. 2004;44(2):134–9.
5. Vogel RA. Measurement of endothelial function by brachial artery flow-mediated vasodilation. Am J Cardiol. 2001;88(2A):31E–4E.

6. Roy-Chaudhury P, Spergel LM, Besarab A, Asif A, Ravani P. Biology of arteriovenous fistula failure. J Nephrol. 2007;20(2):150–63.
7. Wixon CL, Hughes JD, Mills JL. Understanding strategies for the treatment of ischemic steal syndrome after hemodialysis access. J Am Coll Surg. 2000;191(3):301–10.
8. Yeager RA, Moneta GL, Edwards JM, Landry GJ, Taylor Jr LM, McConnell DB, et al. Relationship of hemodialysis access to finger gangrene in patients with end-stage renal disease. J Vasc Surg. 2002;36(2):245–9; discussion 249.
9. Berman SS, Gentile AT, Glickman MH, Mills JL, Hurwitz RL, Westerband A, et al. Distal revascularization-interval ligation for limb salvage and maintenance of dialysis access in ischemic steal syndrome. J Vasc Surg. 1997;26(3):393–402; discussion 402–4.
10. Morsy AH, Kulbaski M, Chen C, Isiklar H, Lumsden AB. Incidence and characteristics of patients with hand ischemia after a hemodialysis access procedure. J Surg Res. 1998;74(1):8–10.
11. Asif A, Leon C, Merrill D, Bhimani B, Ellis R, Ladino M, et al. Arterial steal syndrome: a modest proposal for an old paradigm. Am J Kidney Dis. 2006;48(1):88–97.
12. Zanow J, Krueger U, Reddemann P, Scholz H. Experimental study of hemodynamics in procedures to treat access-related ischemia. J Vasc Surg. 2008;48(6):1559–65.

Chapter 56
Ischemic Monomelic Neuropathy

Gerald A. Beathard

Case Presentation

The patient was a 63-year-old diabetic female who had been on hemodialysis for 1.5 years. At the initiation of her dialysis, a left brachial-cephalic arteriovenous fistula (AVF) was created. Immediately after surgery, the patient experienced severe hand pain, and a diagnosis of ischemic monomelic neuropathy was made at that time. She was immediately taken back to surgery and a proximal arterial inflow (PAI) procedure was performed (Fig. 56.1). At that time, the patient was found to have a marked weakness of all intrinsic muscles of her ipsilateral hand and wrist. She was started on physical therapy and has regained some use of her hand. At the time she was seen, she could move all fingers and almost make a fist.

She continued on dialysis but has experienced recurrent cephalic arch stenosis. This has been treated at 4-month intervals. She was referred to the vascular access facility because of blood flow on dialysis and a hyperpulsatile AVF. Neurological examination of her hand muscles revealed that she had moderate intrinsic muscle atrophy in both hands, and although she was able to move her left wrist and fingers, the range of motion was limited (Figs. 56.2 and 56.3), and there was decreased strength in wrist extension, weakness of thumb opposition, and weakness of both extension and flexion of her fingers. In addition the patient's left hand was cold in comparison to the right. She complained of no current hand pain at rest or on dialysis.

After the patient was prepped and draped, an angiogram was performed (Fig. 56.4). This showed marked stenosis at the cephalic arch and several stenotic lesions in the proximal cephalic vein. These lesions were successfully treated with a 10 × 4 angioplasty balloon (Fig. 56.5).

G.A. Beathard, MD, PhD (✉)
Lifetime Vascular Access, University of Texas Medical Branch,
5135 Holly Terrace, Houston, TX 77056, USA
e-mail: gbeathard@msn.com

© Springer International Publishing AG 2017
A.S. Yevzlin et al. (eds.), *Dialysis Access Cases*,
DOI 10.1007/978-3-319-57500-1_56

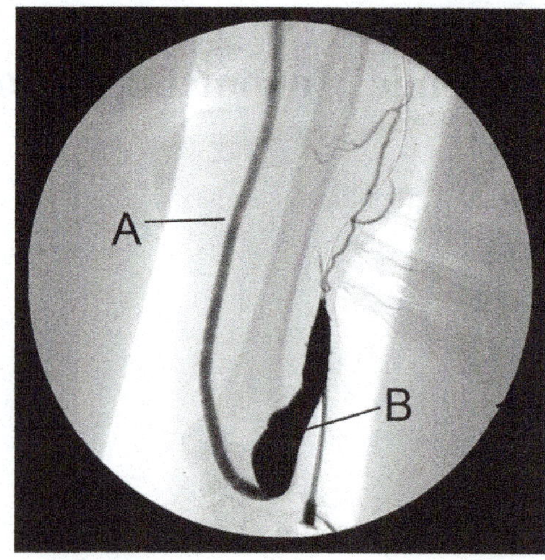

Fig. 56.1 Angiogram of AVF demonstrating PAI relationships. *A* ePTFE extension connecting AVF to axillary vein, *B* lower portion of AVF. *AVF* arteriovenous fistula, *PAI* proximal arterial inflow

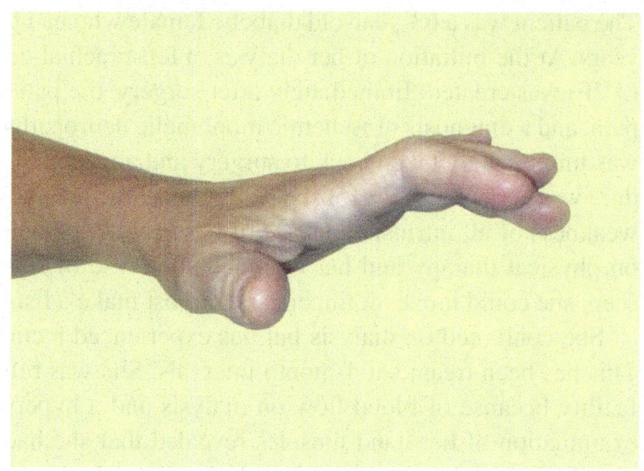

Fig. 56.2 Lateral view of patient's hand showing appearance suggestive of an ulnar nerve palsy (claw hand deformity)

In order to make appropriate plans for future vascular access management in this patient, angiogram of the basilic vein and the images from the original vein mapping procedure were reviewed (Fig. 56.6).

Discussion

This patient represents a typical case of ischemic monomelic neuropathy (IMN). Although the appearance of the patient's hand as shown in Figs. 56.2 and 56.3 suggests an ulnar nerve palsy, neurological testing indicated involvement of all three nerves that supply the hand – the ulnar (weakness of intrinsic muscles of the fingers), the radial (weakness of wrist extension), and median nerves (weakness of

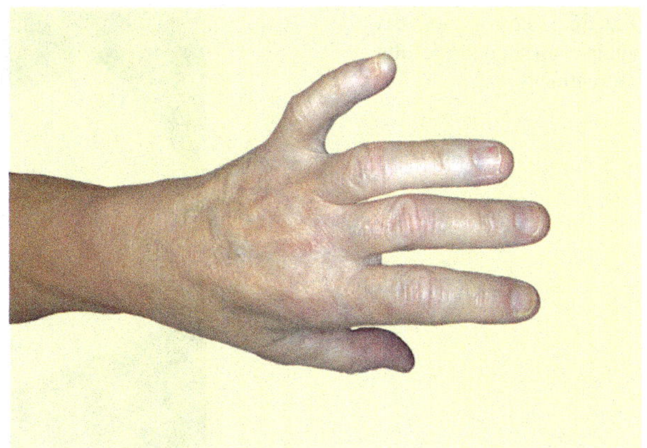

Fig. 56.3 Top view of patient's hand

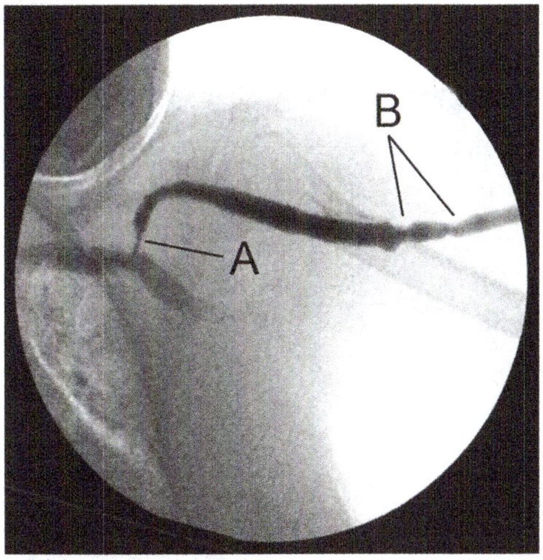

Fig. 56.4 Angiogram of cephalic arch showing stenotic lesions. *A* stenosis near junction with subclavian vein, *B* areas of stenosis within proximal cephalic vein

thumb opposition). Paralysis of a single nerve in the setting of vascular access surgery should exclude the diagnosis of IMN [1] and prompt a search for a local nerve compression secondary to a complication of the surgery [2].

IMN is a condition that, on its surface, appears to be unique: an ischemic injury occurs almost exclusively in diabetics that is selective for one tissue (nerve), but spares other tissues (muscle, skin). The pathogenesis of this phenomenon as it relates to IMN is dependent upon two factors – the characteristics of the microcirculation of the nerve and the comparative metabolic needs of the tissues involved (relative rate of oxygen consumption).

Based upon human anatomical studies, the antecubital area has been shown to be a watershed zone for the *vasae nervorum* perfusing the three nerves that supply the lower arm [3]. With occlusion of the brachial artery at the time of AVF creation, the area of poorest perfusion (and thus maximal damage) is not the distal vascular field,

Fig. 56.5 Post-angioplasty
angiogram showing results
of treatment

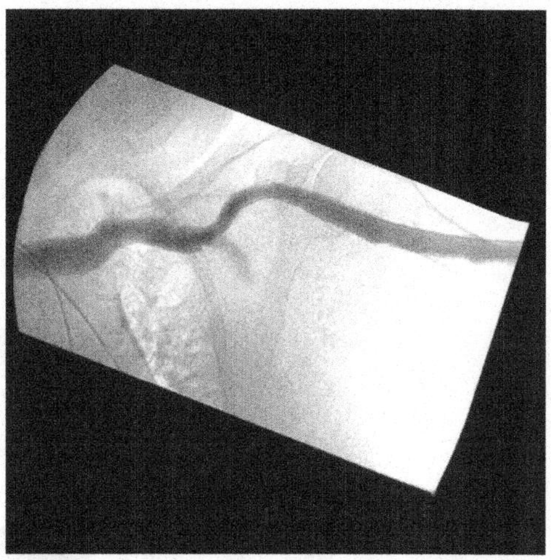

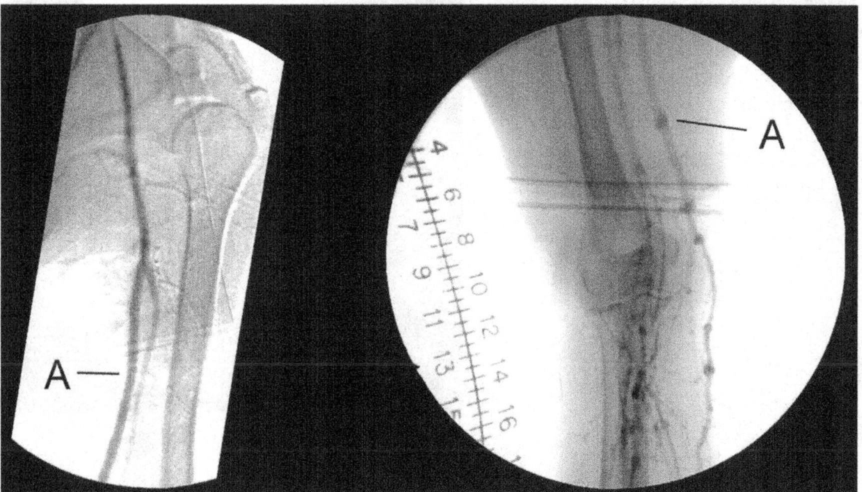

Fig. 56.6 *A* (*left panel*) normal appearance of basilic vein in ipsilateral arm. *A* (*right panel*) normal appearance of basilic vein in contralateral arm from original vascular mapping

but a band of nutrient vessels to the nerve in this watershed zone. Thus, selective nerve damage occurs.

Additionally, the selectivity for neural injury that is seen in IMN may be in part due to a greater metabolic requirement for these peripheral nerves than for other tissues creating a difference in relative infarction thresholds [1].

Early recognition of this complication is crucial. While it is not unusual for a patient to complain of pain following surgery, it is important that the nurse locate the pain when a generic complaint of such a symptom is reported by the patient. Access surgery on the arm should not cause pain in the hand. Severe pain in the hand following a brachial artery-based dialysis access procedure should immediately raise the suspicion of IMF. IMF can generally be easily diagnosed clinically based upon its immediate onset of typical neurologic symptoms and signs involving all three nerves to the hand in the absence of evidence of ischemic injury to other tissues. As soon as IMN is diagnosed, immediate closure of the access is recommended in order to have a chance at preventing severe and irreversible neurologic injury. Even with early access closure, paralysis and pain may be permanent [4–6].

This patient is unusual in that her access was not closed, but a revascularization procedure, PAI, was performed. This procedure is designed to convert the arterial supply of the AV access to a more proximal arterial level by using a small-caliber (4 or 5 mm) expanded polytetrafluoroethylene (ePTFE) graft as a feeder [7, 8]. In doing this the original arterial anastomosis is closed. A length of ePTFE is connected to the distal end of the access and extended up to a new arterial anastomosis with the axillary artery as is shown in Fig. 56.1. The goal is to improve distal perfusion while preserving the access.

In this patient, the question is what should be done at this point. The fact that she is a diabetic with a cold access hand suggests that she is at definite risk for the development of dialysis access steal syndrome (DASS). Given a patient is with this clinical picture, great caution should be exerted in treating a stenotic lesion in the access outflow. Decreasing resistance to flow in the access could precipitate hand ischemia. This patient has had several angioplasties performed without difficulty, however. The presence of the PAI revascularization procedure may be protective in this regard. An option for her recurrent cephalic arch stenosis is to create a new access in the contralateral arm, and the original vascular mapping showed that she had a good basilic vein (Fig. 56.6). This could be very problematic in that she has already been shown to be susceptible to the development of IMN. Should such an event occur involving the opposite hand, she would be rendered very handicapped, unable to use either hands normally. Another alternative for management is outflow relocation, a procedure in which the cephalic vein is swung downward to connect to the basilic/axillary vein, eliminating the cephalic arch; the current studies show that she has a good ipsilateral vein (Fig. 56.6).

Clinical judgment concerning this patient's individual situation is necessary. Continuing to perform an angioplasty at 4-month intervals could very well be her best alternative.

References

1. Miles AM. Vascular steal syndrome and ischaemic monomelic neuropathy: two variants of upper limb ischaemia after haemodialysis vascular access surgery. Nephrol Dial Transplant. 1999;14:297.
2. Reinstein L, Reed WP, Sadler JH, Baugher WH. Peripheral nerve compression by brachial artery-basilic vein vascular access in long-term hemodialysis. Arch Phys Med Rehabil. 1984;65:142.
3. Adams WE. The blood supply of nerves: I. Historical review. J Anat. 1942;76:323.

4. Wytrzes L, Markley HG, Fisher M, Alfred HJ. Brachial neuropathy after brachial artery-antecubital vein shunts for chronic hemodialysis. Neurology. 1987;37:1398.
5. Redfern AB, Zimmerman NB. Neurologic and ischemic complications of upper extremity vascular access for dialysis. J Hand Surg Am. 1995;20:199.
6. Hye RJ, Wolf YG. Ischemic monomelic neuropathy: an under-recognized complication of hemodialysis access. Ann Vasc Surg. 1994;8:578.
7. Zanow J, Petzold K, Petzold M, et al. Flow reduction in high-flow arteriovenous access using intraoperative flow monitoring. J Vasc Surg. 2006;44:1273.
8. Thermann F, Wollert U. Proximalization of the arterial inflow: new treatment of choice in patients with advanced dialysis shunt-associated steal syndrome? Ann Vasc Surg. 2009;23:485.

Index

© Springer International Publishing AG 2017
A.S. Yevzlin et al. (eds.), *Dialysis Access Cases*,
DOI 10.1007/978-3-319-57500-1

Made in the USA
Monee, IL
07 July 2026

56553857R00168